Town a...y
Grow
Trees

Irene Finch

Almond

Longman

Words you may not know
(Use the index too)

abnormal not of the usual kind
bacteria very small live things that can multiply. Some cause decay or disease. Many are helpful
brittle unable to bend much without breaking
conifer tree with cones such as pine
cuttings pieces of twig etc. which form roots if planted
deciduous trees that lose their leaves in winter
fibrous made of fibres
flake flat piece that comes off
fluted column or trunk that has smooth curved furrows and ridges running down it
fungus plant made of cobwebby threads which produce dusty spores, toadstools etc.
furrow a groove—usually cut or split, or dug out
gall a lump on a plant caused by the irritation of a pest
hive home for bees
lacking missing, without something
microscopic tiny, seen only by using a microscope
obvious easily seen
pigment coloured substance
prevailing wind commonest direction for the wind
prune trim off branches
stub short piece of a pencil, branch etc.
vandals people who cause damage
virus tiny live particles even smaller than bacteria (above)
wilt go soft through lack of water

★ shows things worth collecting (see p. 13)

Contents

Trees, buds and spring	4
Girdle scars	5
Leaf series	6
Leaves with wings	7
Young leaves	8
Old and young leaves and their protection	9
Leaf mosaics and use of leaves	10
Evergreens	11
Different kinds of leaves	12
Things that fall from trees	13
How stems grow	14
How stems stop growing	15
Autumn in spring and spring in autumn	16
Twigs into tree trunk	17
Tree size	18
Special ways of growing	19
Story of an oak branch	21
Effects of pruning	22
Flowering trees	24
Small catkins	25
Flower into fruit	26
Pollen	27
Trees without fruit	28
Germination	29
Tree seedlings	30
What is bark?	32
Cork and lenticels	33
Older bark	34
Studying bark	35
Twigs	38
Fairly smooth bark	39
Wrinkles and angle marks	42
Seals and scars	43
Growths and bumps	46
Shaggy and furrowed bark	47
Furrows and scales	50
Other things on bark	51–53
Wood	54
Annual layers	54
Different trees	56
Testing	57
Warping and cracking	58
Rays	59
Plywood and veneers	60
Different places, different plants	61
Roots	62
Nature trails and moving pictures	63
Index	64

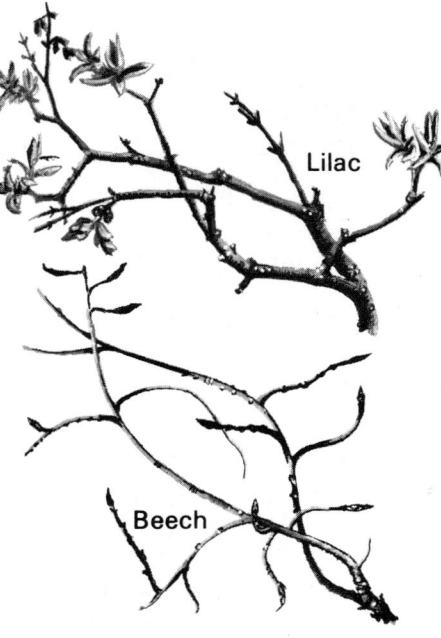

Lilac

Beech

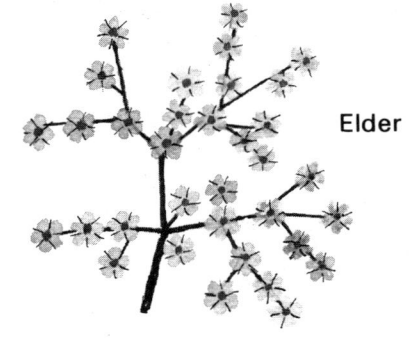

Elder

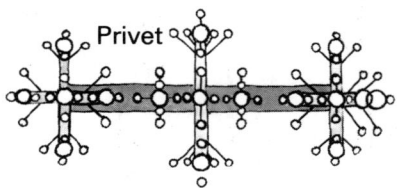

Privet

The *Autumn Trees* book deals with fruits, autumn colours, fallen leaves, damaged leaves, and naming trees. It gives points to look for on each tree and a booklist.

How to study trees

There is so much to see on trees that sometimes it is hard to know where to start. One good way is to choose your own tree, and then, as you go through this book, see what your tree has to show for each chapter. Spring is a good time to start. As you watch your tree through the year, you can collect the different things that fall from it, make measurements and drawings, and find out as much as you can from books. You will then know enough to write a book yourself.

Opening buds

Oak

Winter buds are mostly covered with dark scales to protect them. The spring is an exciting time because the leaves inside these scales begin to grow, so the buds swell up. The scales start to separate and you can see new clean pieces on each scale, that never saw the light before. Some trees start very early in the spring, like privet, but some are very slow, like the ash. A few trees have buds that will open very early if they are brought indoors and 'forced'. Willow and Forsythia can often be forced in January, and birch and lime a little later. Some trees need fairly long hours of daylight before they will start to grow, and warmth alone is not enough. Try some different kinds.

The date of spring

We often say that spring has come when we hear the first cuckoo or see the first crocus, but this depends partly on luck, and it will not work in towns. It would be better to choose a few trees that grow both in the town and in the country, and average out the first date for each. You could fix on the first bright green privet buds, the first almond blossom and so on. Do you think spring comes sooner in the town or the country? Is it different in different parts of the British Isles, on different soils, on hills, in valleys?

Girdle scars

Look for a bud that has grown just a little, so that the bud-scales are ready to come off easily. Pull them off gently, one by one, leaving any that are still firmly joined on. You will find that all the scales are joined to the twig rather close together. When they come off, they leave scars that are very close together and make a ring round the stem, called a girdle scar. This is usually pale green. The rest of the new stem is going to grow on beyond the girdle scar.

If you follow further down the twig, you will probably find another girdle scar, but much darker, usually brown. This is the scar that was made *last* spring, by *last year's* bud-scales. So the bit of twig between the two girdle scars took one year to grow. Look further down the branch and you may find still older girdle scars. Always the piece between two girdles took one year to grow, so you should be able to work out when the bottom girdle scar was formed and the age of the bottom piece of your twig. Look for very old girdle scars that start to split and peel off.

Many of the twigs of cherry and laburnum trees and the shaded twigs on beech trees have the girdle scars very close together. What does this mean? Try to find a twig that is as old as you. Look out for twigs that are making a lot of growth each year and twigs that are making very little. What is the greatest length of growth that you can find for one year on your tree? Do long and short twigs grow in different places? Do flowering twigs grow much? Does a twig grow the same length every year? And are fallen oak twigs (page 13) different from those that stay on?

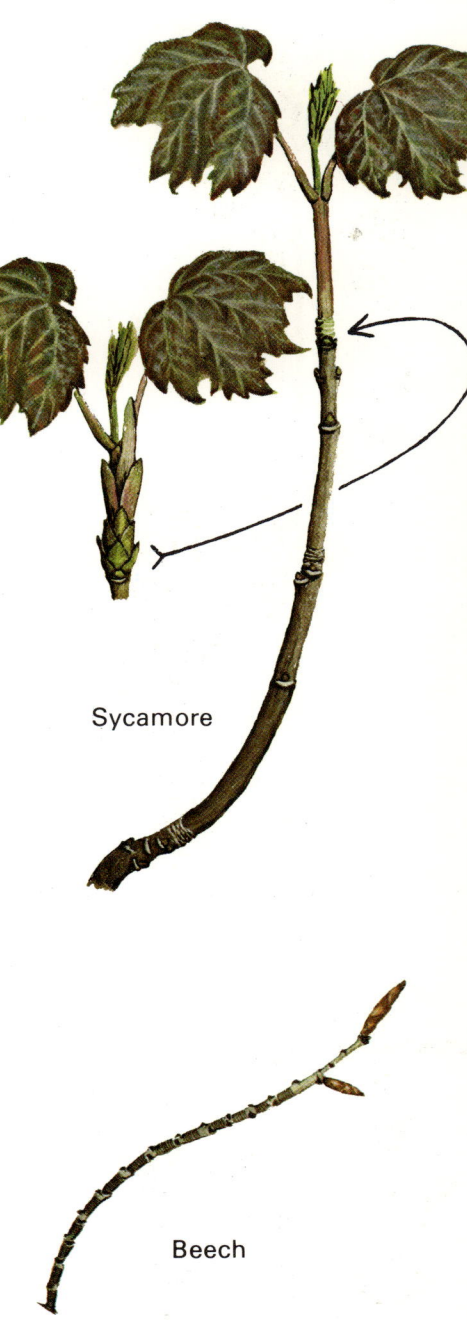

Sycamore

Beech

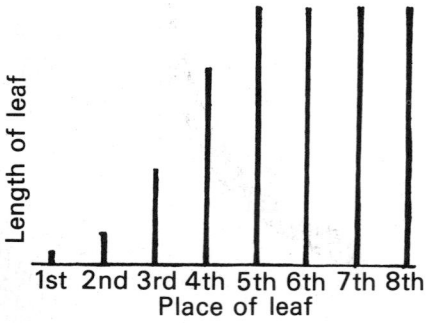

Length of leaf

1st 2nd 3rd 4th 5th 6th 7th 8th
Place of leaf

Leaf series

Look for a spray of privet, over 8 cm long and with the lowest leaves and bud-scales still on it. If you start at the bottom, you find first bud-scales, then small rounded leaves, then bigger ones and gradually more oval and pointed ones, like the usual privet leaf.

Some twigs have small leaves at the top. These are the usual shape but not full-grown (see page 8).

Some privet twigs have every leaf full-grown by May and all the leaves at the top are the same size and dark green. These twigs have a brown bud at the top.

If you take off the leaves one by one and put them in a row, you will have a series. You can press them or keep them under transparent adhesive plastic. You can space them out at equal distances across the page or at the natural distances apart.

These leaves have almost drawn a graph of their lengths. Try putting it on paper, or try a graph of their widths.

Look out for other plants with a good leaf series in spring. Try lilac, aucuba, elder, snowberry and hawthorn. Sometimes the series is different on different shoots. Sometimes you can graph the number of teeth. Some trees, like hornbeam, do not change much in shape, only in size.

Sycamore, ash and flowering currant have a more sudden change, part-way along, from bud-scales into ordinary leaves. Even here you can sometimes find an in-between kind of leaf, like a big scale with a little leaf sitting on top, if you search all the fallen scales. Ash has more of these in-between leaves than most trees.

Plane leaves only change a little along one branch, but you will find different kinds of leaves on different parts of the tree and on trees of different ages. See if you can arrange them in a series. Look at severely pruned trees and very old trees.

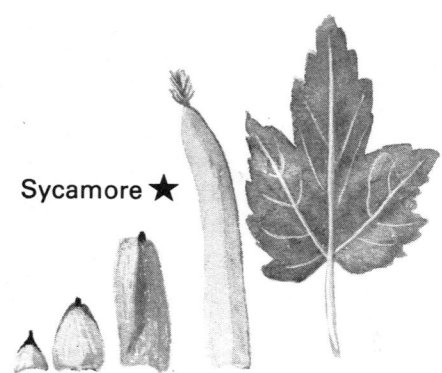

Sycamore ★

Leaves with wings

If you look at a lime twig that has grown about 20 cm, you can find bud-scales alongside each leaf, not just at the bottom of the new stem. They look like little pink or green wings each side of the leaf stalk. You can find them on many other trees, too. Look at the top of the stem and see how the bud is protected. Sometimes each leaf is covered by its own wings but sometimes each pair of wings protects the leaf just above. Horsechestnuts cannot protect every young leaf like this, because all their bud-scales are at the bottom.

At the top of the stem the wings are bigger than the tiny new leaves, so they must have grown faster. But later on the wings stop growing and the leaf goes on and overtakes them, like the tortoise and the hare.

The very bottom scales are usually very tough and make the outside of the bud. They have no leaf with them.

Rose leaves never lose their wings. Nor do the leaves on the long twigs of hawthorn and other trees. But cherry and lime lose their wings quickly, and they leave a scar.

Many leaves have interesting wings. Look at Japanese cherries, hawthorn and quince, if you can. The locust tree and the barberry have turned their wings into thorns.

Plane leaves have the wings joined into a tube that fits round the main bud and protects it. When the time comes for the next leaf and the stem to grow out, they come up through the top of the tube. Look for these tubes just above the place where a really young leaf joins on. The tubes usually fall off quickly, but sometimes they stay on and grow a frilly top.

At the very bottom of the new plane stem there are never proper tubes, only hard dark bud-scales and then little hoods covered with golden fur.

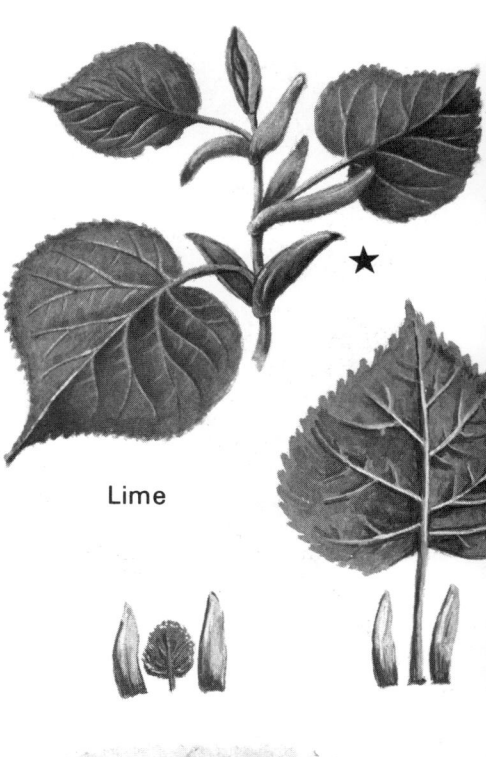

Lime

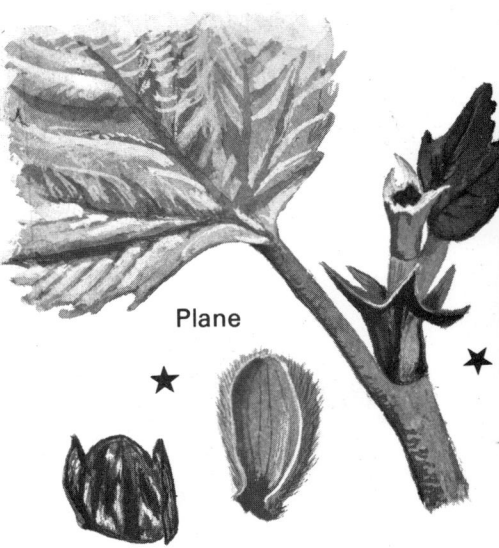

Plane

Young leaves

Folded up flat

Folded under

Rolled

Double upward roll

The leaves are packed very neatly in the bud, usually folded or rolled up. Look for some of the different ways shown here.

If you take a winter bud to pieces, you will find a few tiny folded leaves inside (page 14). Inside these leaves is hidden the tip of the stem, and extra leaves can grow out of this. They are tiny lumps at first, hidden by the other small leaves. During the winter, the bud has scales round it, and does not grow. It is a resting bud. But the buds on spring and summer twigs are growing new leaves all the time. To see how the leaves grow in a bud, take a large bud to pieces: a long lettuce is a good bud to choose.

When the leaves have unfolded from the bud, they still have to grow. If you can, measure the first leaf that gets clear of the bud, without injuring it, and then measure it again regularly, you will see how it grows slowly at first, then faster, finally slowing to a stop when it is full size. If you can, try this on a leaf that is growing outdoors.

Many twigs have pale leaves near the tip that get smaller and smaller and, at the very tip, a bud. The small leaves, laid out, will show you the stages of growth: they will all grow to their full size in time (see second paragraph on page 6).

Some leaves are poisonous.

Pleated

Young leaves are delicate and cannot develop a tough skin till they are full-grown. They are often torn, damaged, or broken off, and they dry up easily. If you leave them without water, they wilt fast, and even if they stand in water they can die in a warm, dry room. It is best to keep them cool and out of the sun. At night they can be sprinkled with water and covered.

Some young leaves are partly transparent when held against the light, and the veins show clearly.

Older leaves grow a waxy skin that helps to stop them from drying up, but it is much better on some leaves than on others. The whole leaf and the veins get thicker and tougher as they grow older, much better for leaf rubbings and for pressing. But most people do not like tough old leaves of lettuce in a salad, and if you give caterpillars a choice, they will usually eat the younger leaves. Even young lime leaves were once used for salad, and you often see wood pigeons eat them. These young leaves also contain more body-building protein than the older ones. Perhaps this is why they often turn brown when you press them.

Older leaves have more green matter (chlorophyll), and if you look along a twig with leaves of all ages you can sometimes see a whole range of colours. See how many leaf colours you can find on one tree. You can always tell a twig is still growing if there are pale leaves at the tip.

These pictures show some of the ways that young leaves use to save them from drying up, once they come out of the bud-scales. Many stay rolled or folded for a time, as on the opposite page. We do not really know if the red colour helps, but it is quite common. Look at your trees for methods of protection and different colours.

Old leaves and young leaves

Curving down

Stays folded

Fluff

Folding down

Hanging down
Yellow fluff

Protecting young leaves from drying up

Curving, red

Leaf mosaics

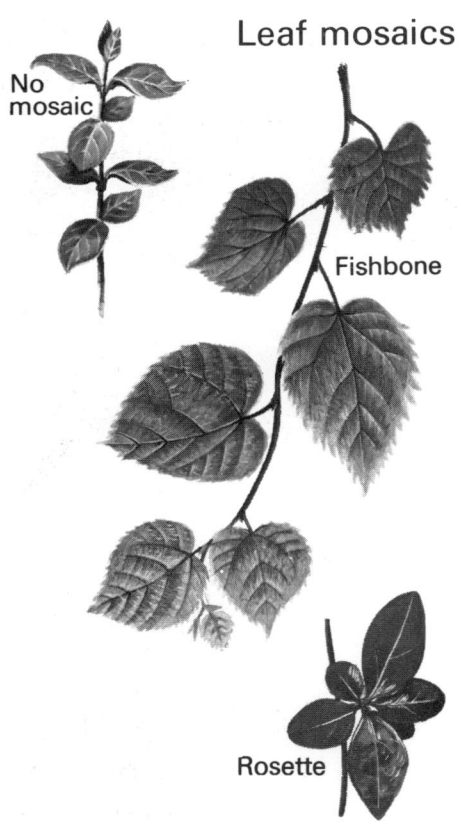

No mosaic

Fishbone

Rosette

The use of leaves

Plate mosaic

High up on trees or bushes, the leaves may stick out all round the stems, or else hang down and flutter in the wind, and they are not always flat. It does not matter if they shade each other a bit, for there is plenty of light.

But the leaves on lower branches often place themselves in a flat plate, so that they do not overlap, but fill each little space like a mosaic. Trees with a good mosaic give shelter from the rain. If you take a photograph or use a light meter you can also see how they cut off the light.

Some trees have the leaves growing alternately, and these make a neat fishbone mosaic. Low sprays of privet are usually short, with the leaves in a rosette, but sometimes privet makes a fishbone mosaic, although it does not have alternate leaves. A careful look will show how.

Young sycamore leaves move about a lot. Try picking a horizontal branch with a plate mosaic and putting it *upright* in water for a day. Or lay a small potted tree on its side.

See how many rosettes, fishbones and plate mosaics you can find on different trees.

This is quite complicated to understand. The first point is that leaves catch the light, and all leaves place themselves to catch the light (see mosaics above).

You can see how important leaves are to the plant if you pull the leaves *off* one groundsel plant and leave them *on* another, and see how they both grow. The leaves make a difference, for the plant with leaves grows bigger and forms flowers. Leaves get most of the food for the plant. Roots also get some food, but only the minerals and water.

Green leaves can catch the sunlight and use it to make food out of air and water. Then the food is used to make flowers, buds and more leaves. Green colour helps: see page 12.

Evergreens

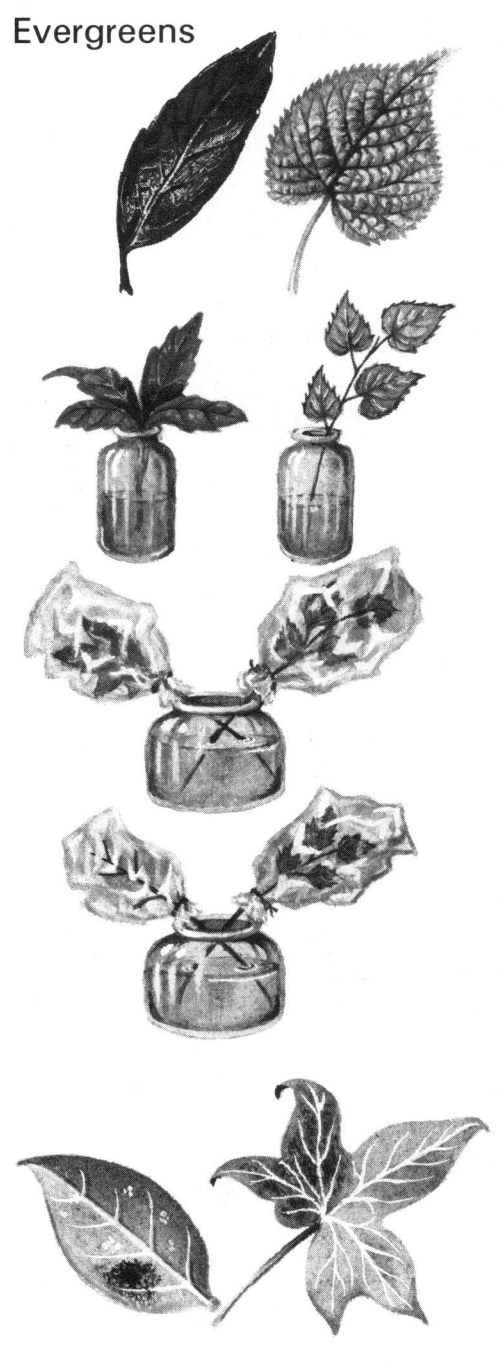

Evergreens do not lose their leaves in autumn, and you can understand them better if you test some leaves in spring and summer.

1. Pick some evergreen and some deciduous leaves and keep them for a few days in a dry place, watching them.
2. Pick an evergreen and a deciduous branch with about equal leaf area. Put each in water in a good light. See how much they drink.
3. Do the same with two more, but tie a small plastic bag over each and weigh how much water they give off into the bag, in a week. You can do this even better on the tree.
4. Try 2 or 3 again, but this time take all the leaves off the deciduous branch, to make it something like the twig in winter.

These tests will show you which kind of leaf loses more water. The evergreen leaves are covered with wax. We use the wax from the tropical carnauba palm leaves for floor polish. Try putting wax polish on leaves in test 1.

Leaves always lose water but the roots usually take in enough to make up the loss. In winter, when it is cold, the roots cannot take in water, even if there is plenty of water about. There are some things *you* cannot do when *your* hands are very cold, too. So in winter it is dangerous for trees to lose much water because they cannot replace it. If there are high winds the loss of water can be very serious. Evergreens keep the water in with thick wax. Deciduous trees do it by losing leaves. Some plant salesmen spray evergreen plants with plastic, to protect them in their first winter.

Evergreen leaves may stay on for several years, as you can see from the girdle scars. See if you can find their season for falling. Privet loses some leaves in autumn but the young ones grow tough and stay on. How long do they last?

Different kinds of leaves to look for

Full-grown leaves are usually flat green plates, but not always.

Some cultivated trees have red leaves or green leaves with white parts. These are all unusual *varieties*, and the original wild plants were green. Copper beeches and red-leaved plums have red leaves all the time, but the wild plant is only red in places, or if there is very strong sun. The red varieties still have some green pigment, as you can see by boiling some water and dropping a young leaf in.

Golden privet seems to have lost some of the green, and variegated leaves have lost it in patches. But all these trees with leaves blotched or bordered in white or yellow are abnormal, usually with a virus disease (not infectious to us). They grow less strongly than normal, just as a striped piece of Tradescantia will grow less strongly than a green piece from the same plant. They would soon die out if they were left to compete with fully green, wild plants.

Leaves are not all symmetrical, i.e. they will not all make equal halves if you fold them down the mid-vein.

Many leaves are not flat, but wavy. They may have too much leaf for the mid-vein, so that they go into waves. These often make good plaster casts. For very flat leaves, look in shady places. Flat leaves may change. Lilac leaves at the top of the bush will roll up in summer, and lime leaves hang down in a heat wave. Many other leaves hang down at different angles, helping you to recognise them.

Leaf stalks are in *Autumn Trees* (page 17) and leaf wings on page 7 of this book.

As well as leaf series, some plants have occasional abnormal leaves, like a four-leaf-clover, or a laburnum. Privet and weigela have some shoots with leaves in triplets, and strongly pruned limes may have leaves with large teeth or a funnel shape.

Look along the street and round the park after wind or rain; many different things fall from trees as the spring and summer go by. Try to see which tree they come from.

In this book, the pictures with black stars will give you some ideas of things to look for, starting with buds, bud scales and flowers. By June the first yellow leaves start to fall. Plane fruits fall in spring though they form the autumn before. Willow, elm and poplar fruits fall in June because the flowers are early and the fruits grow fast.

If you keep a diary you can compare with other years and places. Keep specimens, but make sketches of things that wither and change colour. If you record the rainy weather and wind speed (Beaufort scale) you may find what causes the falls.

Dead twigs often break and fall in gales, but the break is a danger, for it may allow decay to enter the tree. A few trees have a better way: they cast off unwanted twigs neatly, and seal the scar. You can always tell these cast twigs by the properly shaped piece at the base, and the way they leave behind a deep pit with a rim. Look under oak, poplar and willow trees. Black poplar even casts twigs with live leaves, in early summer.

In spring, find a small side-twig of oak on a branch, and peel it. The wood at the base is very weak and will sag in as it dries. This is the place for the break. But a cork seal is prepared before the twig actually comes off. Search for twigs just ready to fall off in summer, and make the final break yourself. Only side branches are cast and only some of these, and only up to a certain size. Try to find the biggest possible cast twig. Look out for *freshly* cast twigs. Were the freshly cast twigs growing less well than those that stay on? Girdle scars show how much growth they made last year.

Things that fall from trees

Falling twigs

Oak

How stems grow

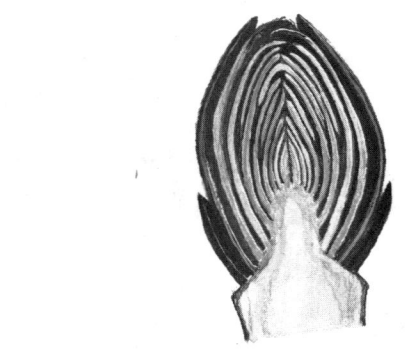

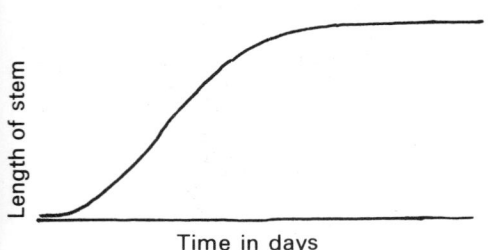

Time in days

This picture shows a bud cut open, with its short stem and very crowded leaves. The stem will get longer as the bud grows, so the leaves will be spaced out.

Choose a bud growing outdoors and try to measure the length of the stem every day. You will have to guess where the stem tip comes, or judge from a bud you have cut open. If you start with a really tight bud it will probably not move for some time, then it will start very slowly, speeding up after a time. Then the graph goes up more steeply.

If you keep on with your graph of stem length it will gradually slow down. Many people give up at this point, thinking something has gone wrong. But it is normal for the growth curve to flatten over till it becomes quite horizontal, meaning 'full stop'. The whole growth curve is now complete, in an S-shape. A bean plant has a growth curve of similar shape, so has a single leaf, and so have most animals, including human beings.

To find which part of a privet twig is growing, dip a fine pen in Indian ink and make tiny blobs, very close together, all along the stem. Then see if any will grow wider apart.

Another way to find which part is growing is to leave a privet twig without water till it just starts to wilt. The part that is still growing will sag over, but the fully-grown part is rather woody, and stays stiff. Of course this method is no use for plants like tulips that *never* get woody.

Some buds never open. Look at them carefully, for they may turn into galls.

This graph was made by marking the position of each leaf, as well as the stem tip. You can see that the plant grows by adding on extra sections, and extra leaves. Animals do not grow like this, adding on extra joints or ears.

How stems stop growing and prepare for winter

Growing / Stopped growing

Lime / Lilac / Plane

If you measure twigs each day, you can see if they have stopped growing. But you can often tell more quickly by a glance at the leaves. Growing twigs are making pale young leaves at the tip. Stems that have stopped growing will lose their pale tips. Look for stems that have stopped growing in summer. And in shady places, twigs often stop growing even in April.

Sometimes you can find a flower bud at the end of the twig when it stops growing. And in early August some birch shoots stop growing to make next year's catkins. But usually, when a twig stops growing, it does not form a flower, but only a small *winter* bud, covered with scales. It may be a bit small and green, but it is ready for winter. The stem has stopped growing and has made bud-scales instead of leaves, though the weather is still summery. We do not know exactly what causes growth to stop, and you will find that some branches and trees stop before others. Look in sun and shade, and at pruned and unpruned trees. By August the side-buds usually swell up too and are ready for winter.

Lime and plane trees do a strange thing when they stop growing. The top of the stem does not turn into a winter bud, but it withers and finally drops off. The picture shows the poor little lime tip, but the plane tip falls off in its tube (page 7).

When lilac stems halt, the top bud usually withers, but stays in its place. But instead of the next bud down the stem taking over the lead, there are always two buds, with the withered one between. What will happen when these grow out? Does the real top bud ever survive?

Autumn in spring, and spring in autumn

We often call autumn the fall, but the old leaves of evergreen trees sometimes fall in *spring*. Lime trees lose a few bright yellow leaves in June. They were badly lit and not doing well.

In smoky towns, or in dry places and dry years, leaves may turn brown and fall in August. One branch of an elm tree may turn brown in summer because of a fungus disease that blocks the water-carrying tubes of the twigs.

Trees sometimes lose their leaves because caterpillars eat them bare. But as you saw on page 15, they already have resting buds ready for next winter. These grow out, as if it were spring already, and replace the leaves that were lost. If caterpillars eat the leaves in May, the tree can have a second spring in June. I have seen horsechestnut trees lose their leaves in a dry town square in August. More rain came in September, and this made some of the flower buds open, with pale green leaves. But other trees in the next street, that had been well watered all summer, were turning to autumn colours and dropping conkers on the ground.

Privet bushes can have a new spring during the summer if the bush is clipped and more light is let in. You can often find two or even three leaf series on a privet spray because the final resting bud was not allowed to rest. On other trees you often see a branch broken and the side bud growing out with pale leaves. The oak tree is famous for its lammas shoots, which can make the whole tree look as if spring had come in August.

When these things happen the girdle scars cannot tell the right age, and there are thin, extra, annual rings. See if you can find things happening out of season.

Sometimes, after leaves are lost or pruned, new shoots grow out of unexpected places, such as trunks and thick boughs. They come from tiny dormant (sleeping) buds which you can sometimes find on twigs. These dormant buds do not grow unless an accident wakes them.

Twigs into tree trunk

First summer

First winter

Second summer

Second winter

Third summer

These pictures show how a twig can change into a branch in three years. The twig in its first summer has a row of buds down the stem, one to each leaf, and these are left in the winter. Next spring they all grow out, each with a girdle scar at the bottom. As time goes on, there are more and more branches, side branches, twigs, leaves and buds, and to carry all this weight the main branch has to get thicker and thicker. You could make a film of this (page 63). If you could continue with these drawings for another forty or fifty years, the main branch would turn into a trunk about a foot across and a seedling would grow into a full-sized tree. You can find the age of each twig and branch by counting the girdle scars. All the twigs in these drawings are greenish in their first year, light brown in their second year and dark brown in their third year. Would you say a real two-year-old twig was twice as thick as a one-year-old? Does it have to carry three times the weight when it is three years old? Is it four times as strong when it is four years old? You can test this by taking twigs of equal length and putting them across a gap. Find what weight you have to hang on them to break them. Use the same piece of tape to hang the weights each time. Does a twig twice as old carry twice the leaf area?

Tree size

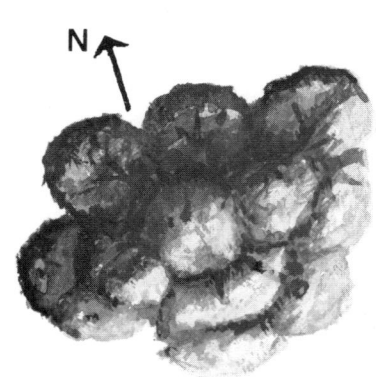

Special trees can be measured exactly. The circumference of the trunk, divided by three, will give its diameter roughly. The volume of usable wood is worked out roughly by finding a quarter of the circumference, in metres, multiplying it by itself and then by the height of the trunk. Do you think this gives a fair measure? The height of the trunk and the height of the tree can be measured in different ways, which can be used to check each other. One is to measure a friend's height and let him stand against the tree. Then hold up a ruler to estimate how many friends you would need, stood on top of one another, to make the height of the tree.

Another way is to measure the length of the shadow of a 1 metre stick and then measure the length of the tree shadow. The spread of the tree in all directions is also worth measuring as accurately as you can, from the trunk. Walk to the edge of the branches looking up to check where you are, or getting someone else to check. Then measure the distance in. Find your compass direction from the trunk and write that down too. Then try other directions, at least eight in all, and draw them all on a map of the tree. Trees rarely have the same spread all round. Find the directions of the prevailing wind and work out which side gets most light. Do these explain any lopsidedness?

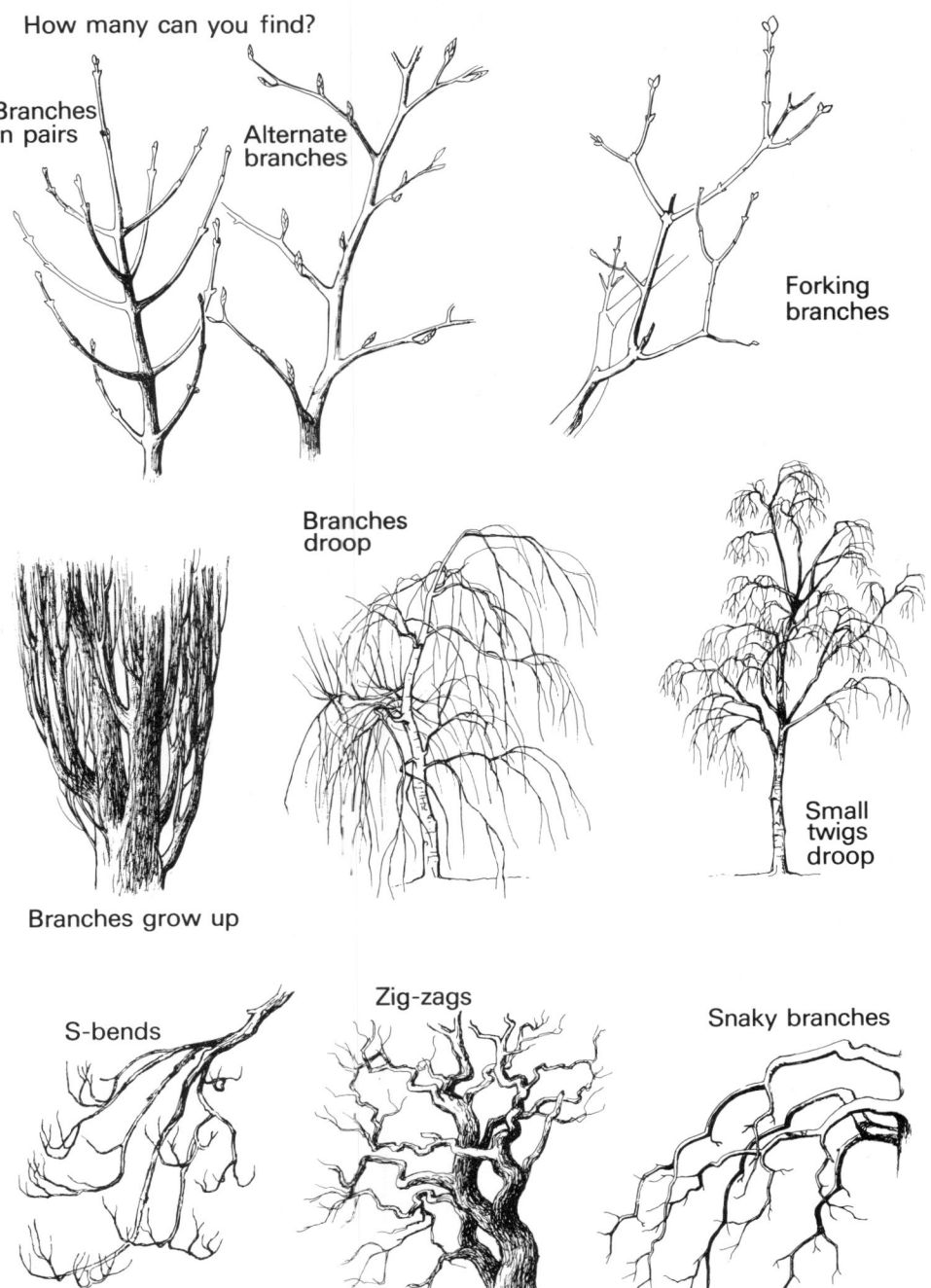

Special ways of growing

Story of an oak branch

Effects of pruning

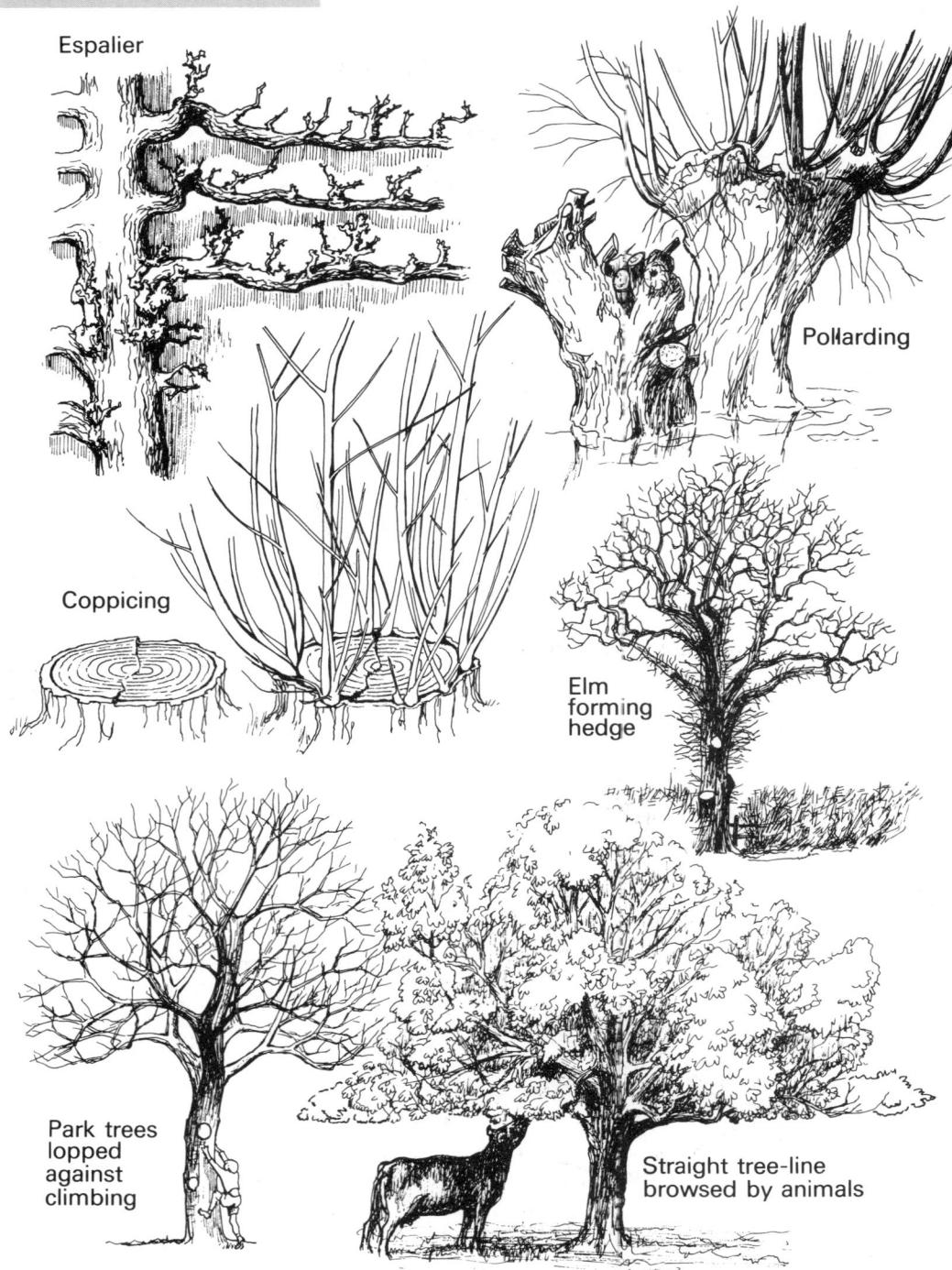

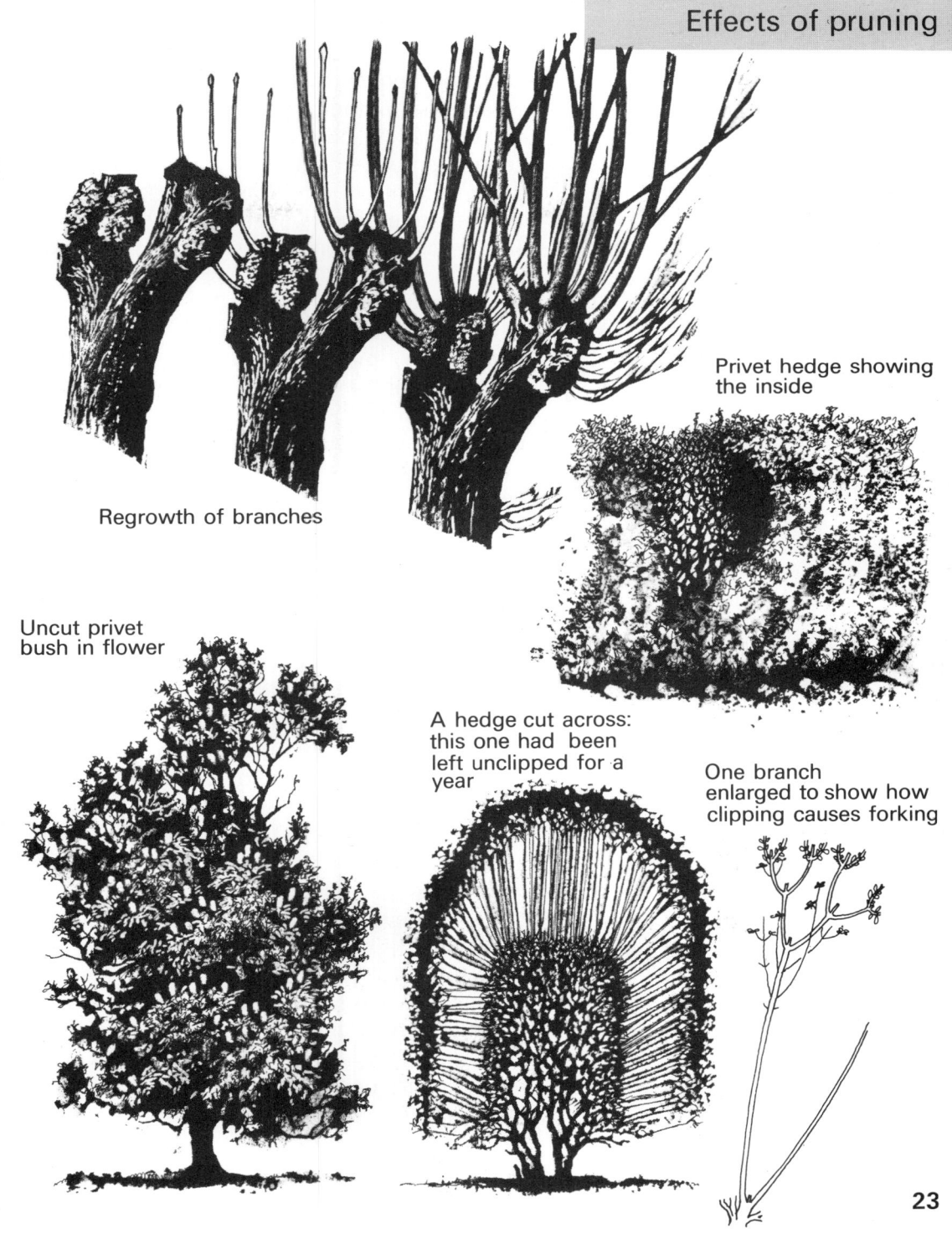

Flowering trees

Blackthorn
Pear
Norway Maple
Sycamore

Many trees have ordinary flowers, and you can find: the *sepals*, that cover up the bud; the *petals*, that attract insects; the *stamens*, that make a yellow dust called pollen; and in the centre, or underneath, the *pistil*, which will grow into a fruit.

Many tree flowers have scent and nectar to attract insects. Some have patches of colour called honey guides, that show the way to the centre of the flower.

Other flowers are smaller and greenish, like the Norway maple in this picture, but bees still come for the nectar and pollen. The nectar comes from the green pad under the stamens. Lime flowers are also rather small and greenish but they have a powerful scent and plentiful nectar, attracting many bees. Sycamore flowers attract bees but are smaller still. They are arranged in a long bunch or *catkin* that hangs down. Some of the small flowers on this very big catkin have no pistil, only stamens. Some big horsechestnut flowers also lack pistils.

Rhododendrons and honeysuckles have a *tube* of petals, instead of separate petals. Laburnum flowers have a complicated shape like a pea-flower. Try to see these.

If your trees have big flowers, you can make out the parts and the shapes and colours. In sunny weather, see which insects visit them, and what they do. Some insects have long tongues to get nectar, but moths only come out to honeysuckle at night. Bees collect the pollen in little pads on their legs, and then take it back to the hive, to feed to their grubs.

Many small flowers are arranged in beautiful patterns, which you can see on page 3. Privet flowers are in twos and fours, making a regular pattern if you map them. Elder has a clear pattern when it is in bud. Look for other patterns made by small flowers.

Most forest trees have greenish flowers, tiny and hard to see, arranged in catkins. Catkins often come out early before the leaves are full-grown. The bees do not come for the pollen, and it floats out in dusty clouds.

Look for birch twigs with long catkins at the ends of the twigs, and leaves just coming out. The long catkins have about fifty tiny flowers, each with a bunch of stamens, but these fall off and turn to nothing.

But a little later, amongst the leaves, you can find a smaller kind of catkin, bright green and standing upright. It is made of tiny green flowers with little red pistils on each.

Plane trees have yellow-green catkins shaped like a ball, and made of stamens, and also, higher on the tree, a smaller number of bright red ones with pistils.

Oak has long yellowish catkins with stamens, and very tiny ones, like pins, with pistils.

Elm and ash flowers come out before any leaves, and they look like reddish fluff with stamens and pistils. Elm is often out in February but the flowers are very high up.

Willow and poplar catkins are silky at first, but later on you can see stamens or pistils between the silky hairs. Willow catkins are unusual because they are visited by bees, collecting early pollen and nectar.

Look out for catkins on trees: most people miss them.

Small catkins

Birch

Plane

Oak

Elm

Ash

Aucuba

Flower into fruit: growth series

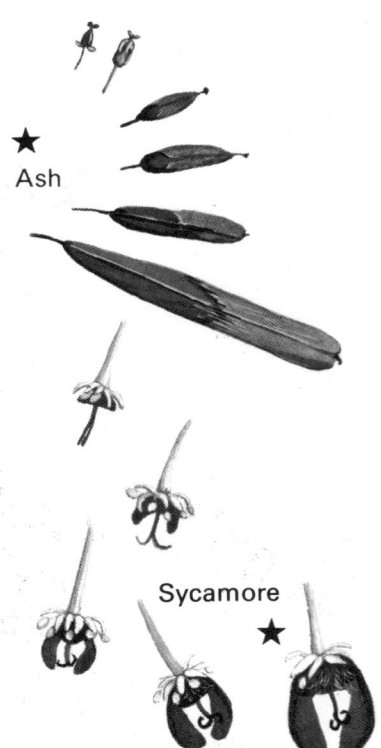

Ash

Sycamore

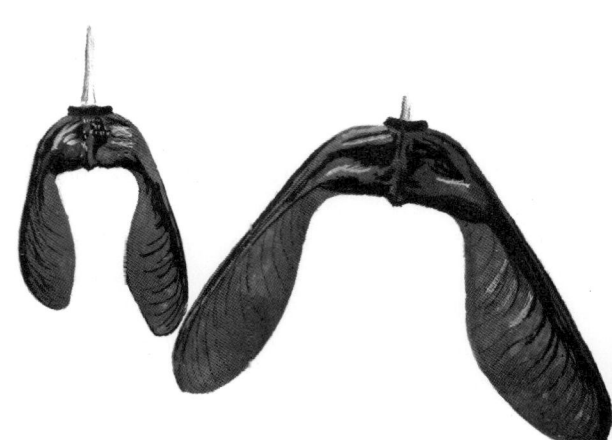

Look for the chance to see how a flower changes into a fruit. On low bushes, like aucuba and rose, you can watch the small ovary, the main part of the pistil, grow bigger and bigger, and change in colour, while all the rest of the flower dies away. Inside the ovary the seeds grow and darken.

Tall trees like elm, sycamore and ash are often just as easy to study, for some flowers and fruits are falling all the time, and if you collect them you will get a series, which can easily be preserved by pressing or under transparent plastic. You can see from the picture how the wing gradually grows out, and the stigma at the top part of the pistil withers.

There is more about these fruits on page 20 of the *Autumn Trees* book, including something about the fruits that come down in summer.

Pollen

Bees collect pollen to feed their grubs, but it is also useful to the plants themselves, and the bees do not keep it all.

If the right kind of pollen is placed on a pistil, it can grow into a fruit. The pollen sticks to a knob or tuft on the top of the pistil, which may be greenish or red. This knob is called the stigma. If it catches even a little pollen of the right kind, the half-formed seeds (ovules) inside the pistil can grow into proper seeds. You can test this out if you can find a plant of aucuba (page 28) or white campion that has pistils and no stamens. If you keep the bees away with muslin, you can put pollen on some of the stigmas and not on the others. This can be tried with snapdragon flowers, but get rid of their stamens before they open.

Many trees have a dusty pollen that blows in the wind, and muslin bags would not keep it out. You can see this dusty pollen if you keep young oak or birch catkins in a still room or a small plastic bag for a day, and then shake the branch. The wind carries this kind of pollen to the stigmas.

Other trees have sticky pollen that is carried by insects while they visit the flowers. Bees are so good at this job that many orchard keepers put a hive in the orchard. Other insects carry pollen, but they often mix up the different kinds or visit rubbish dumps in-between.

Sometimes you see a gap down a spray of lupin fruits because the weather was bad when these flowers were opened, and no bees were about to carry pollen. But the gap at the end of horsechestnut sprays and sycamore sprays is caused by flowers without proper pistils.

Flowers produce millions and millions of grains of pollen that are lost in the wind or eaten by bee grubs and other insects. Only a small number of these pollen grains manage to land on a stigma of the right kind.

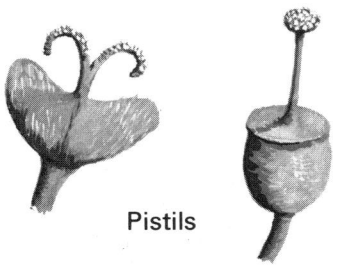

Pistils

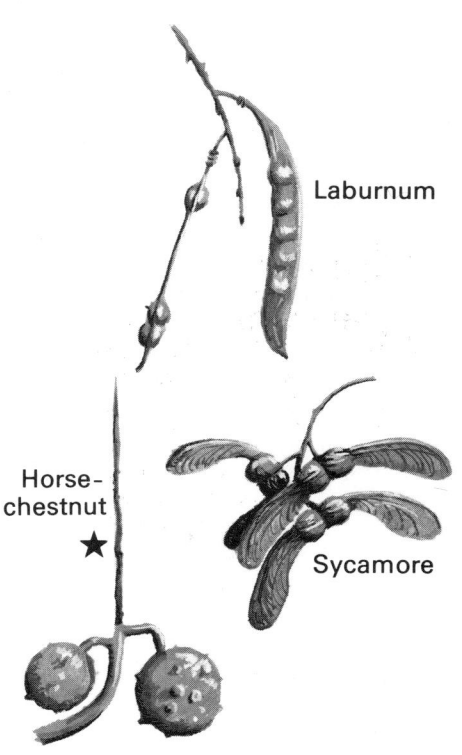

Laburnum

Horse-chestnut

Sycamore

27

Trees without fruit

Aucuba

Holly

Tree-ferns and much-pruned trees never flower, but some trees have flowers and no fruit.

These may be varieties where the stamens and pistils are turned into extra petals. These double-flowered trees are often planted in parks to stop children from doing damage in a search for fruits, e.g. double cherry and horse-chestnut.

Other trees have fruit on some bushes but not on others. Look at holly and aucuba, which carry the berries right round to the spring, and see if the berried trees are bigger or different in any way. Aucuba bushes flower very early and have interesting flowers. Bushes without berries have flowers like those on the left, with fat-stalked stamens and no pistils. Watch what happens to these. On the other kind of bush there are different flowers, with pistils but no stamens. The pistil is partly under the flower.

This arrangement prevents the bush from being pollinated by its own pollen (self-pollination). You will find that many plants have ways of preventing self-pollination, and this is useful to them but complicated to explain.

If the pollen bush and the seed bush are about the same size, the pollen bush usually has more flowers, and a longer flowering season. Can you see any point in this?

Holly trees and some others have the same kind of separation, but there may be useless small stamens or pistils on the other sort of tree. The trees without berries often grow bigger.

When the aucuba was first brought in from Japan, it did not fruit though the flowers had pistils. Can you explain this?

Seeds have a small plant inside, with two rather fat seed-leaves. There are pictures of some of them in *Autumn Trees*. When the spring comes, many of these seeds will start to grow, but some will wait another year before they sprout.

First the small root grows out. Then the seed leaves usually lift themselves out of the seed coats, to unfold and spread out. The seed leaves are green, but rather thick, because there is food in them. The tiny bud between them starts to grow and soon it sprouts the first ordinary leaves. These are thin, but they may still not be quite the usual shape.

The little plant is called a seedling while it still has the seed leaves, but in the end they will drop off. At the end of the year, unless the plant is an evergreen, all the leaves will fall, and you are left with a tiny winter tree, with its own winter buds, but very thin and hard to find.

Next spring the main bud will unfold from its bud-scales. Some of the bud-scales may have a tiny leaf on top. It is not a seedling any more, but a one-year-old tree. But it may have, growing round it, the new seedlings of the next year. You can see which is which if you look for the pair of seed leaves. Some of the plants may be two years old, or more, and you can check this by the girdle scars.

A few seeds do not germinate this way. Acorns, for example, do not lift the seed leaves above ground. Perhaps they are too heavy. If you have seen peas and broad beans growing, you will remember that they also keep the seed leaves underground.

Germination

Sycamore

First winter

Second spring

Oak

29

Tree seedlings

The seedling trees on these two pages can all be found in spring, but some start earlier than others. Some are found near their parent tree, but some are good travellers, and sometimes you can measure how far they went. Methods of travel are in *Autumn Trees*, pages 20–29.

You often find the seed-coats on the ground, near the seedlings, but usually there is no lumpy seed inside. They have lifted the seed leaves, with the food inside them, into the air. The seed leaves are quite different from the ordinary leaves, but these pictures will help you with the names. Try to find some very young seedlings and watch them unfold. Some will grow in flower pots for a few years, as small trees.

The oak and horsechestnut seedlings are often found in late spring, near their parent tree. Brush the soil away gently and you will find that the acorn and conker are still there, still supplying food. This is because the seed leaves have stayed underground.

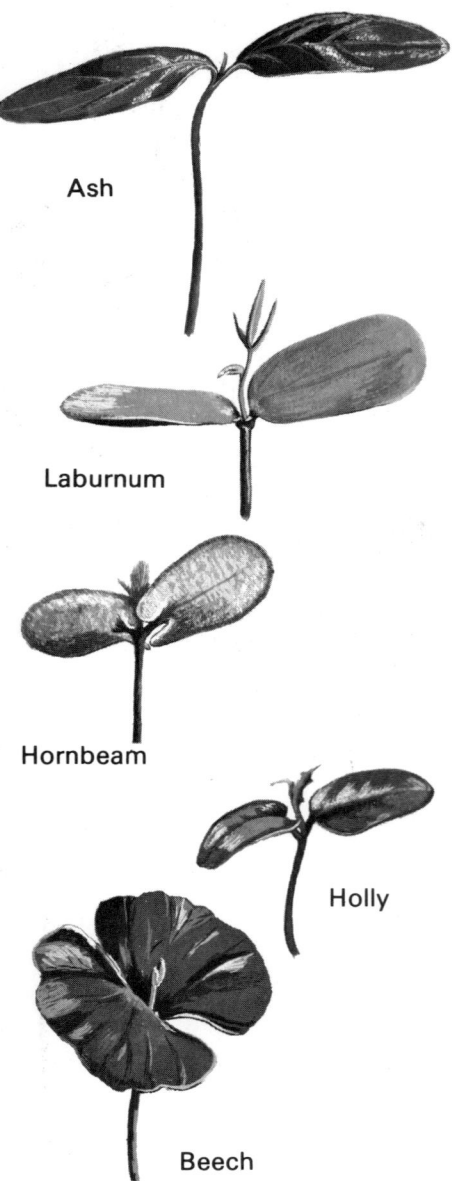

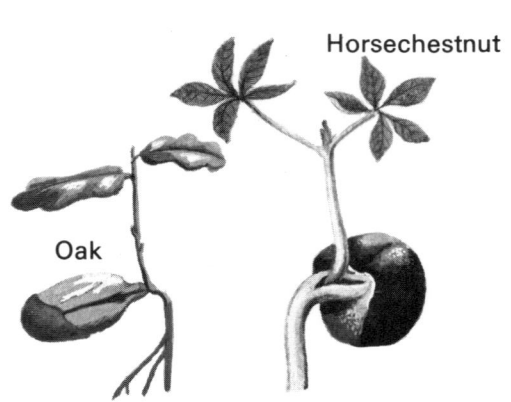

The seedlings at the bottom of this page are less often seen. The birch seedlings are common and the seeds travel well, but they are very small. You may be able to find them more easily later in the year, perhaps in June, when they have grown a few leaves. The likeliest places are mossy patches in the forest, and bare, damp, sunny places everywhere. Plane seedlings usually do badly in this country, but they are common beside the river Seine in Paris. In London they are mostly found during the year that follows a very good summer, on mossy patches and in gratings. Like lime seedlings, they seem to be completely missing in most years. Yew, pine, and spruce seedlings will usually be found in places where these trees grow. Look out for seedlings near any tree you are studying.

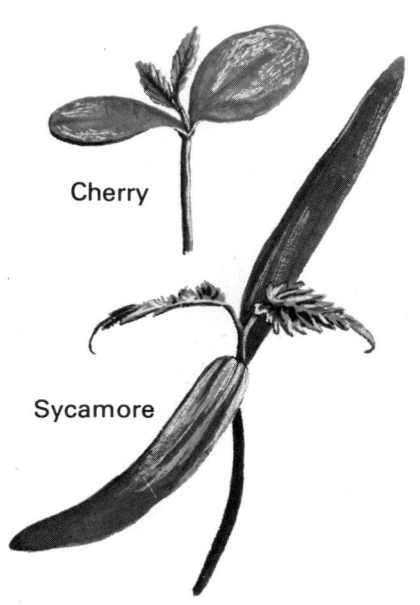

Cherry

Sycamore

Hawthorn

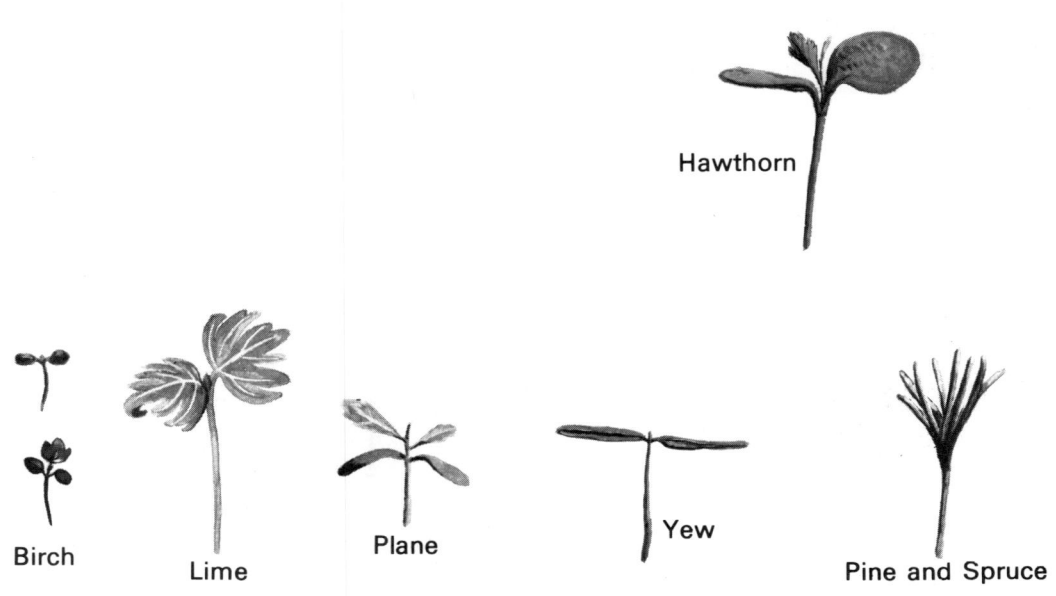

Birch

Lime

Plane

Yew

Pine and Spruce

What is bark?

Branches and trunks were once twigs, so we start in early spring with a small twig that has been through the winter and has buds growing out. Very new twigs have no bark. Peel off the bark: it comes away easily in spring, but seals itself on by late summer.

Your twig peels neatly because it has a soft layer that breaks easily. This is a growing layer, called a cambium. The inner part is the wood (page 54).

The bark you have peeled off is mostly made of bast. It has lattices of fibres, rather like linen threads, running down the twig. Bast is usually strong and flexible. Lime bast was once used to weave cloth to make mats. Thin strips of elm bark, soaked well, were perhaps used to tie together the wooden poles of Stone Age huts. The American Indians often used elm for canoes instead.

The drawing shows lime bast under a lens. The little diamonds are not good fibre and can be removed by pounding, soaking and washing. Raffia is the bast of a foreign plant. Linen is the bast of flax, which is easy to grow from seed. Nettles were also made into very good cloth by early men in this country. They were picked at the end of summer and the woody part pounded out.

The inner part of the bast has soft tubes that carry sugar to the tree roots. The tree will die if they are harmed, and new ones have to grow each year.

Bast often gets stony and stiff with age. Beech bark has little stony pieces in it, and the bast of most trees becomes hard on the bigger branches, with the fibres less plain. But here and there you can still see fibres, especially on lime, honeysuckle and locust trunks.

Cork and lenticels

Young twigs have a thin semi-transparent layer of cork, which you can peel off in spring because it has a cork cambium, just underneath. If you wash it, you will see that it is waxy and shiny, but is not darkened or wetted by water. It is almost air- and water-proof, and it keeps the water in the twig. A twig without cork will soon wither. So will a potato.

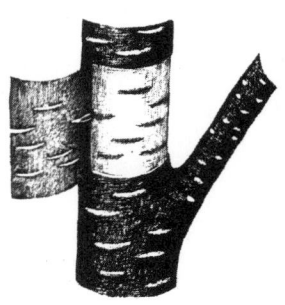

The thin piece of cork you peeled off will have a few holes in it, partly filled with powder. On a fresh twig these look like small bumps with powder coming out, and they are called *lenticels*. The powder is a loose kind of cork that lets a little water out and a little air in, in summer. In winter, lenticels are sealed behind.

Under the cork is a soft layer, often green, but not very important. Then comes the fibrous bast.

Bottle-corks are made from a foreign, evergreen, oak. Its cork grows unusually thick, and in regular layers, each annual layer sticking on to the next. This cork can be peeled off neatly at its own *cork* cambium, to get the cork, and the bast beneath is not injured. (This cannot be done with any other trees, for other trees do not have a complete layer of cork cambium, and cannot be peeled without damage to the bast and the tree.)

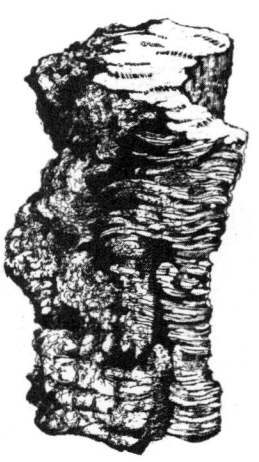

Cork keeps liquids in bottles, and the water in the tree. But lenticels run right through from outside to inside and make a weak place. Look at a bottle cork to see what the manufacturer has done about this. You can scrape some of the lenticel powder out and look at it with a microscope.

Cork resists heat and protects the tree to some extent against hot sun and sharp frost. Test a cork mat and see.

Cork also contains tannin that protects the tree against fungus-rot, but tannin is also found in the other parts of the bark, in the heartwood (page 55) and in oak marble galls. Oak-bark tannin can be used to preserve leather. It does this by changing its chemical nature.

When tannin touches moist iron it makes black ink. You can often see a streak from the iron nails in a pale coloured fence. Ink was once made from oak-gall tannin and a solution containing iron. Try rubbing a nail on the bark of your twig. If the cork layer round the tree trunk is broken, disease and decay can attack (page 44).

Older bark of ordinary trees

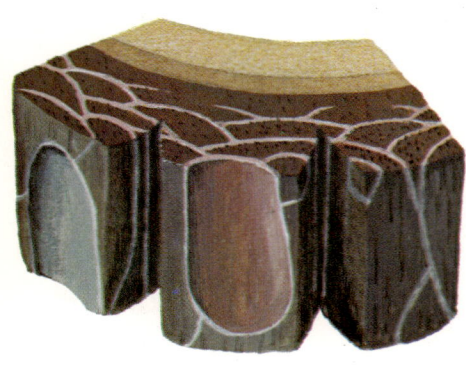

Enlarged

Fresh cut   Colour after
cork   exposure

At the end of the summer, the old bark is getting too tight, and splitting. Look then.

The cork is in thin layers in ordinary bark, not thick chunks like cork oak. Young white birch and cherry trees have several thin layers of cork close together, white in birch, pinkish in most trees (page 39). The layers show well at the edges of the lenticels. As the tree gets older and thicker, the cork cannot stretch much, and soon splits and peels off. It is replaced by new layers from below, made of a tough kind of cork. It may split into rings, but it is strong against horizontal pulls. Indians used it for canoes and boxes. The lenticels lengthen sideways as the trunk thickens.

Many twigs have much thinner layers of cork, and sometimes the red or green of the twig shows through. This peels off in tiny silvery pieces, seen only with a lens. Silvery sycamore trunks and greenish laburnum trunks give off these fine flakes that leaves the bark almost smooth (page 38).

After a time, however, many trees grow the new cork layers much further in, among the bast fibres. Often there are thick 'scoops' of bast in between the cork layers, as you can see in this section of an elm trunk. When the old bark gets too small, and splits, you can see bits of cork and also bits of bast. Pinkish cork goes silvery with exposure (page 39). It shows up well on elm and some oaks. Some walnuts have white cork. These thick layers may flake off in time or they may stay and make great thick layers with furrows. If this kind of bark is peeled off, the inner bast and the inner layers of cork are damaged, the tree may die.

Every tree has different kinds of bark at different ages. Sun and shade, wet and dry, grazing and cutting can also affect bark. Young 'white birch' trees have brown trunks. Very old and very young plane trees do not have flaking bark. Elm is often similar to oak, and horse-chestnut to sycamore. This causes problems, so it is safest to use the leaves for naming trees, checking with flowers and fruit, and then look at the bark to see what you can discover for yourself. Washing often helps.

Look at the bark of your chosen tree all over. Start with the smallest twigs and see the colour, texture and markings. Then look at larger and larger twigs and branches till you find something different in the bark. Look at the new kind carefully, perhaps making drawings. How big are the twigs when they change? Then go on to bigger and bigger branches and the trunk itself. Sometimes you will need to visit another park for the biggest sizes. Field glasses will help with high branches. See if pruned trees, or trees in shady or windy places, are different. Try to watch your tree when it is actually splitting or flaking the bark. The best time for most trees is August and September. Notice the colour of the clean new bark and see if it changes later.

Sometimes tree trunk bark is difficult to make out at first. It may be injured or perhaps doing two things at once, like growing anglemarks and splitting them up. You may have to search to find places where a tree is doing something quite clearly, such as making a branch seal. When you find something really clear, and worth seeing, show it to other people. Mark it on a map or nature trail, or make some kind of record.

Studying bark

Birch

The coloured pages starting at page 38 have pictures of different kinds of bark, to look for and to understand. There are many beautiful patterns and you can make bark rubbings like the leaf prints in the *Autumn Trees* book. A large piece of heel ball, from the shoe-mender, or a really thick wax crayon, are best, and you need a large piece of thin, tough paper and someone to help you hold it steady when you wrap it round the tree trunk. Rub all over the paper, evenly and steadily, with the wax, and leave no gaps between one stroke and the next. Do not rush straight up to the tree and take a bark print of the first bit you see. Look out for a really interesting piece to do. Most bark prints need careful trimming and mounting afterwards, to show them off well. They need not all be square: try long thin ones and other shapes. Sometimes the bark prints will cut up well to different shapes to make fish, birds and so on. The 'wax-resist' method gives a different kind of print. Use ordinary candlewax instead of heel ball and then, when you get home, flood the paper with watery paint, which will not stick on the wax.

Coloured drawings can be even better. Faded bark exposes brighter layers when it is cast, and a spray of water brightens many colours, but dulls silver. You can also make a thick layer of papier-mâché (flour and wet newspaper) on a board, and scrape *out* the pattern you want, or make a 'drawing' in thick paste. For plaster casts, press clay or Plasticine against the bark. Lift it off to use as a mould. Lay the mould on a table, print up, and build a wall round the edge as shown in *Autumn Trees*. Then pour in plaster.

If you look through the next few pages up to page 53 you will find coloured pictures of some of the things you can find on bark. These pictures will give you ideas of things to look for. Or you may find some interesting bark for yourself, and then you can look through the pictures, to see if it is in this book. Behind each picture you will find a description in words.

Cork shreds and lattices
The twig with the buds is oak, and it has thickened, so that its coat of cork is too small. This cork is shredding off in wispy pieces, quite hard to see, and a new smooth layer of cork has already formed underneath. The wider twig is splitting its old cork into a lattice which is so fine that you have to look carefully to find the splits.

Splits through lenticels
This is a poplar twig in its first year, but it has grown too big for its bark, and has split it. The splits show clearly and run nearly parallel, through the lenticels. The girdle scar at the bottom is splitting, too. All this is because the twig is getting thicker and the cork coat cannot stretch.

Silvery flakes
This twig is fairly thick but it still looks smooth because the silvery old cork (now too small) is coming off in tiny flakes all the time. You may need a lens to see them. Some trees go on doing this till the trunk is quite thick. Even the girdle scars flake away.

Scale insects
These are often the size of young lenticels, or bigger. They may be whitish or brownish, and there are different shapes. They lie flat on the bark, but you may be able to get them off with the point of a needle. Underneath there is often a little grub, sucking the sap of the tree. It may be

Lenticels and leaf scars
The leaf scar is just under the bud, and has the marks of the broken veins. The lenticels are the tiny, scattered holes in the cork, with powdered cork coming out. This twig has cork that is so thin, you can see through it. If you peel it off, you will see that the twig underneath is green where it was shaded, but reddish where the sun hit it. A lenticel sometimes looks like a small mouth when you use a lens on it (see also page 33). Look for lenticels on different twigs. Privet often has white ones, in summer. Do extra lenticels appear on thicker twigs? Or are they wider apart than on young twigs? Are they any bigger on bigger twigs?

Twig scar and girdle scar
These neat pits are only found on a few trees, but are common on oak and poplar. They mark the place where a twig was neatly cast off, as on page 13. Look for the fallen twigs, under the tree. Sometimes there is a leaf scar below the pit.

The girdle scar at the bottom of the twig was made when the bud scales fell off in spring, as on page 5. Some of the bud scales had little buds by them, very close together.

---

bright orange in colour, but it is hidden by the scale, which is made partly of its own cast skins. You can buy a leaflet from H.M. Stationery Office about scale insects.

Twigs

Lenticels and leaf scars

Cork shreds and lattices

Splits through lenticels

Twig scar and girdle scar

Scale insects

Silvery flakes

Fairly smooth bark

Peeling bark
and
long lenticels

Silvery stripes

Silvery lattice

Lenticels
multiplying

Diamonds

Smooth
silver

39

Silvery stripes

This is a lime trunk, but lime trees vary a lot. The silvery stripes are made of thin cork layers that peel smooth. The dark shallow furrows, in between the silver, have broken ends of cork and bast fibres. In late summer, they develop pinkish splits. Most of the lenticels are in the dark furrows, and these also split, and show pink, in late summer.

Silvery lattice

This is a hornbeam trunk. The silvery network is smooth cork that flakes off. The dark part has broken edges of cork and a brown split in summer. The bast is thicker under the silver parts, and this makes ridges.

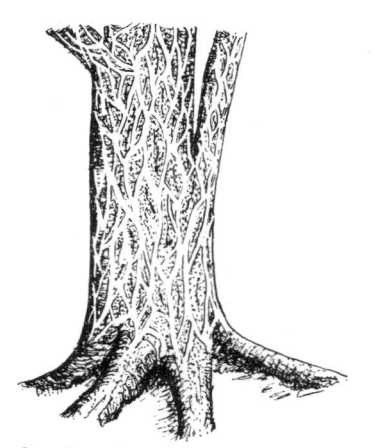

Old hornbeam

Smooth silver

This is the trunk of a small beech tree. The silvery cork flakes off smooth, with only thin layers of bast between, even on old trees. Here and there in the bast there are stony pieces that make bumps and start brown splits in August. These small splits later turn silver and smooth down with the rest.

Peeling bark and long lenticels

The lenticels have stretched as the branch grew thicker, so they are different shapes on different sized branches. They seem to keep pace with the growth of the trunk. These are birch lenticels, and cherry (page 35) has similar ones, but light brown in colour. The lenticels often split into two lips. You can scrape out the powder from these big lenticels with your fingernail and look at it with a microscope. Then you can often see the thin layers of cork at the edge. These gradually peel off in strips that run round the trunk. If you can peel off really thin strips of birch bark, so thin that they cannot injure the tree, you will find they are waxy and can be used to light a fire. Some lime trees have big pinkish lenticels but they do not grow long, and they turn grey with the weather. But many seem to lose their lenticels on the bigger branches and trunk.

Lenticels multiplying

This old cherry trunk now has a row of lenticels where there was only one long one before. The layered cork is peeling off, and the bast underneath it is not yet splitting into furrows.

Diamonds

These diamonds seem to form round lenticels and are rather rough. The rest of the cork is smooth. Soon the diamonds will run together and turn into furrows. Grey poplar has dark diamonds on white cork, walnut has dark brown on lighter brown, and laburnum has brown diamonds on greeny-brown.

Anglemarks

These are flatter anglemarks than those on the other side of the page, and they cannot really be called Chinese beards. These flat ones are found round branches that have stopped growing, either because the branch is dead or because it has been cut off. So there is often a seal (page 44) in the middle. When a branch is dead or cut off, it is not growing, but the trunk still grows, and this causes the Chinese beard to flatten out gradually. If the branch came off a long time ago and was small, the anglemark is very flat, and the seal is small. The forester is pleased, because in that time a lot of solid wood grows over a small branch knot, and there will be a lot of sound timber.

Buds and furrows

The green branch is a thin lime branch, and you can see that the old buds that belonged to the girdle scar (page 38) are still there, but they are small and will probably never grow. They are now among the branch wrinkles. The pink and white trunk is a silver birch with small buds beginning to grow out from the branch wrinkles. Perhaps these will form a burr (page 46). At the bottom of the trunk the bark is starting to split into furrows. The splits seem to start at the anglemarks. Perhaps this is a weak place.

Splitting anglemarks

This anglemark is old and is beginning to split up. In the end it may flake right off.

Wrinkles

The brown trunk is hawthorn, and it has wrinkles under the branch. Perhaps the branch is gradually sagging. The silvery trunk is holly, and the branches are very thick where they join on. There are wrinkles all round the branch. They look like girdle scars, but the girdle scars disappeared long ago. The wrinkles may form because, once the branch thickens, the bark round the base of the twig is too big, so it falls into folds.

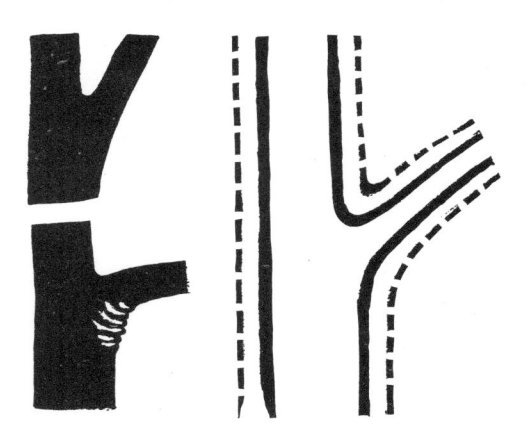

Anglemarks

These marks are often called Chinese beards. They are sets of wrinkles with a peculiar shape that droops each side. They are probably caused like the circular ones above, but this anglemark shape is more common. You can see these marks on many young trees with smooth bark. Beech and sycamore are very good while they are fairly young. Birch trees have very black anglemarks that really do look like a forked beard.

Wrinkles and anglemarks

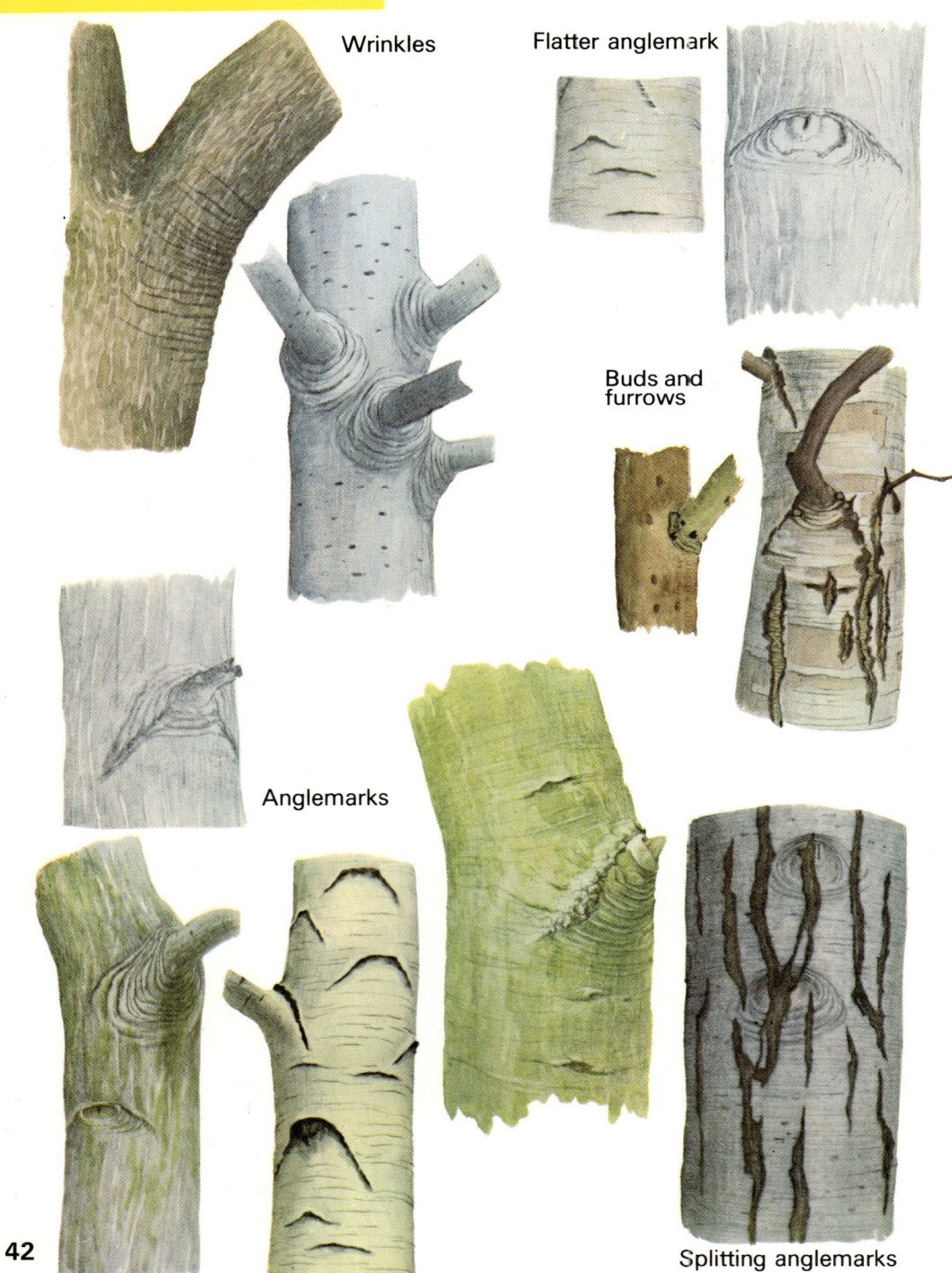

Wrinkles

Flatter anglemark

Buds and furrows

Anglemarks

Splitting anglemarks

Seals and scars

Lopping

Stubs and strips

Holes

Seals

Cuts

Seal patterns

43

Stubs and strips

This branch has been lopped badly, leaving a stub, and this will not seal well. It will collect water and decay. A good 'tree doctor' cuts branches flat against the tree and covers with mould-resistant paint. Lower down in the picture there is a long thin wound being sealed over, but the bark may never meet and the tree may never heal properly. The wood is exposed and will split. This kind of damage can be caused by frost, sun, lightning, deer, or vandals.

Holes

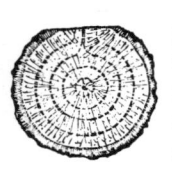

This branch could not be sealed properly by the tree. It faced up, so it caught a lot of water and started to rot inside, making a hole. The water comes out through a hole lower down, and the rot is spreading through the tree. Dead leaves are decaying in the hole and there are even a few pond animals living there.

Cuts

This plane tree trunk has been damaged with small cuts and scratches. At the top you can see how the bark turned brown round a scratch. Under here, a bark roll (see other column of this page) will be developing, and will turn into a seal. You will see this roll, like those on the right of the picture, when the dead brown bark flakes off. The seals are bright brown at first, but fade to grey. You can often find these marks on trees in towns, but these damaged trees do not have such big flakes as other trees, nor such bright colours. The bark seems to stay on a long time, and turns dull grey.

Lopping

Low branches are often cut off park trees. The top drawing is a fresh cut and you can see the reddish wood with dark heartwood, and the pinkish, furrowed bark. The bottom drawing is an older cut, and the bark and wood have turned grey.

Seals

In the top coloured drawing, a rim of new bark is starting to grow over the cut. Inside this roll there is new wood and bast which you can see in this drawing of a seal cut across.

The seal in the lower drawing is several years old and has grown right across the wound, meeting in the middle. You can see lenticels on it and also a star, made of wrinkles. This is because the roll had to fall in folds as it grew into the small space in the middle. You can make a Plasticine model to see this. The finished seal will keep out disease and rot.

Seal patterns

The top seal has a good set of branch wrinkles round it still. You can tell the size of the buried branch by the width of the seal itself. The lower seals are on a rough-barked oak tree: one is beginning to split. It is called an oak rose and may be large or small.

Burrs and lumps

The burrs on the lime tree in the top drawing are caused by the way the tree was pruned. The tree keeps on sending out twigs from this spot, and they are always cut off, so there are hundreds of short pieces of twigs, side by side. The next drawing is an elm burr covered in twigs. It is a natural burr not caused by pruning, but caused when many small twigs start and then die back. Some people say that the buds that make a burr all grow from an old seal (page 43), but probably they grow from the branch wrinkles (page 42) around a seal. The wood inside the burr has little rings all over it, caused by the little twigs cut across. This picture is curly maple or 'birds eye' maple, which is used for expensive furniture veneers.

The other lumps have no twigs and are different. The big one was caused by bacteria, which infect the tree but do not seem to do much harm. Wood boring beetles often lay their eggs in these swellings and when they are full-grown the grubs bore holes to get out.

The smaller lumps, on a beech tree, were caused when a short stub was sealed over as on page 43. The bark is quite smooth now and has some sound wood underneath it, but there is probably a decay spot deep inside, where the old stub is left.

Suckers

This is an elm tree which is growing little shoots, some distance away from the trunk. The shoots are called suckers and they are growing up from the roots. Sometimes the leaves are not quite the usual shape at first

Bare trunks and twiggy trunks

Plane trees and some others usually have very clean trunks, with perhaps an odd sprout at the bottom. But elm trees are noted for the tiny twigs that grow all over the trunk and main branches.

Fluting

Deep flutes like this are common on old hornbeams and are sometimes found on other trees. The dips may be caused when a branch takes food from the bit of trunk just below it, and stops it thickening properly.

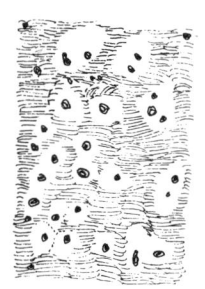

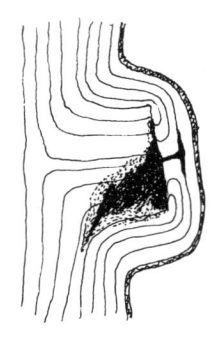

and the stem may be very thick, but in time each sucker can form a new plant. They may push up in the middle of a flower bed or an asphalt path. Groves of elm are formed like this, or rows of elm in a hedge. Plum and cherry often do the same.

Growths and bumps

Bare trunks and twiggy trunks

Burrs and lumps

Suckers

Fluting

46

Shaggy and furrowed bark

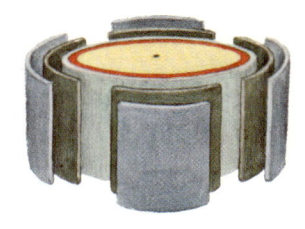

Shaggy bark

How ridges and furrows are formed

Bark with ridges and furrows

47

How ridges and furrows are formed
These pictures show how the bark of a young tree splits and new bark forms underneath. Next year, the bark is bigger still, so that layer of bark splits, and yet another, bigger one, must form below. The splits are all more or less in the same place, and the old layers stick on, so in time there are many layers, with narrower ones on the outside. You can make a model of this with layers of corrugated card. The smooth part of the corrugated card can represent cork and the corrugations can represent bast. The real layers are often less regular 'scoops' (pages 34, 49).

Bark with ridges and furrows
The bark with deep ridges and furrows is poplar. In the summer you can often see a new split in the depths of each furrow. The trunk with spiral furrows is sweet chestnut, but sweet chestnut does not always spiral, and many other trees can spiral. The bark with short furrows, making a sort of lattice, is ash. The last one is a kind of walnut.

Shaggy bark
When honeysuckle or lilac bark gets too small, it comes off in shreds, as on the small trunk in the picture. These shreds give the trunk a very shaggy look, but the birds often come along and tug off shreds to make nests. Here is a piece that was

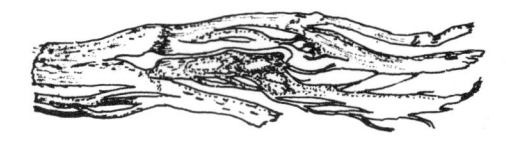

woven into a nest. You can see it was stretched into a lattice before it finally split, like a piece of 'expanded metal'. You can make this kind of lattice with crepe paper if you cut slits in it like this and gradually pull it out sideways. Try long slits and short ones, in different patterns.

The bigger trunk in the picture is a locust tree. It looks a bit shaggy because the different layers all split different ways and partly stick together. This bark is very soft.

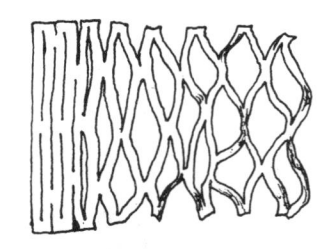

Thin flakes
This is a plane tree at the end of the summer. Most of the bark is greenish grey, because the cork is thin and the green live layers show through A patch is dying and turning brown, because cork is forming under it, and it cannot get food and water any more. The cork forms 2—3 mm under the surface, in the bast. The next drawing shows this dead piece has flaked off. The new bark underneath is clean and yellowish. It soon turns greenish but may go grey if it gets

Ridges and furrows of oak bark

This bark was formed as on page 47, but it has been battered by the weather, and keeps losing layers from the outsides of the ridges, leaving a flattish top. When a piece comes off, the bark underneath is rich brown, but it soon fades to grey. If you look closely at the top of a ridge you can see some silvery flakes of cork, not always quite flat, and the rougher, ridged and fibrous bast. The ridges also have some cross-breaks.

Lacy lime bark

This is rather like the oak, but the silvery flakes of cork are lacy looking. In some furrows, and among the lace there are pinkish lenticels, turning grey. If you look closely at the bast, which is not silvery, you will see a fibrous lattice in places. The silvery lace does not show so well when wet.

Square mosaic of pear bark

The furrows here are broken up by regular cross-breaks into squarish pieces, like a mosaic. They are rather flat-topped because the outer layers keep peeling off. The bark breaks into the mosaic, and peels off very neatly, along special weak splitting places.

Scoops with furrows

The first kind, which looks rather like oak, is really elm. Rather thick, rounded pieces keep peeling off the ridges. If you look in the furrows or if you can see a trunk cut across, you will see the layers of cork and bast quite easily because the bast layers are quite thick. The cork does not grow as complete layers (cylinders) under the old bast, but in little rounded scoops, (page 34) which peel off when they are ready. Try some. They break fairly neatly at the cork.

The second kind is pine. The bast is reddish and the cork layers of a young tree are pale scoops. The scoops will peel off and leave bright reddish bast underneath. As the tree gets older, the flakes mostly stay on and split into furrows. Very old flakes turn to grey (lower picture).

Thick peeling flakes

Sycamore and horsechestnut have bark like this. It is smooth and greyish for a long time (see pages 39, 40). But when the tree is older, it forms a layer of cork deep in the bast. The bark then begins to come off in flakes, breaking at the new cork layer, which is bright brown at first. In time the bark will fade to grey, and later still it will flake again. Look for seals and scars flaking off.

---

really old without flaking off. Water, sprayed on, will brighten all these colours. Look for flakes of bark that have come off. Sometimes they have ridges on the outside, showing where flakes came off the time before.

The shapes often remind you of animals and other things, or give you ideas for patterns. If you draw round a piece of bark and then move it to another part of the paper, you can start a repeating pattern. Then try turning the bark over.

Furrows and scales

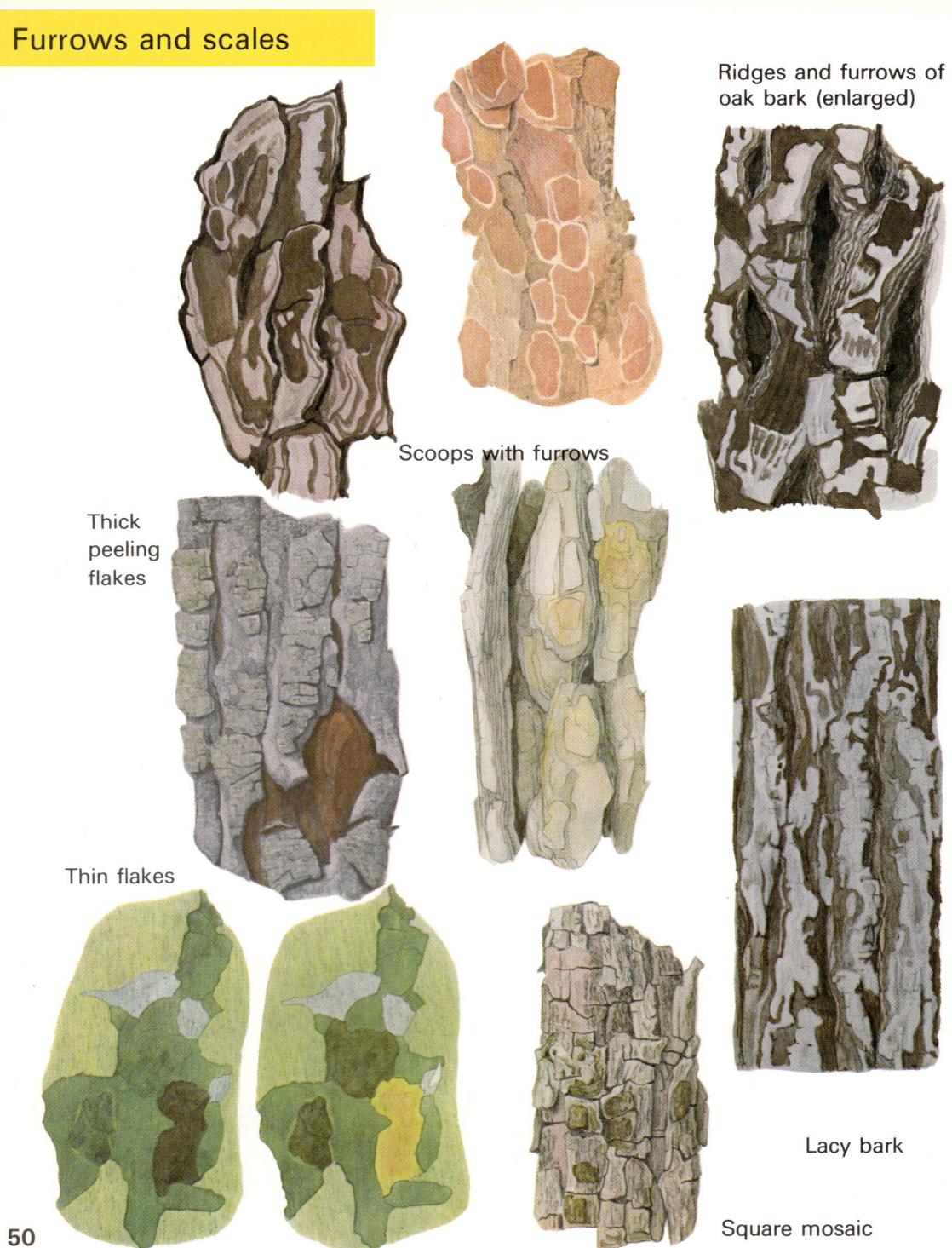

Ridges and furrows of oak bark (enlarged)

Scoops with furrows

Thick peeling flakes

Thin flakes

Lacy bark

Square mosaic

50

Other things on bark

Green flat patches

Resin

Lichens that are not flat

Birds

Streaks

Witches' brooms

51

Resin

These treacly drops ooze out of cherry and pine trees at the end of the summer, if they are injured or split. They protect the injured tree against disease and may also trap insects. The drops contain turpentine which evaporates or can be extracted. When the turpentine has gone you are left with hard resin. Nowadays we can also make resins in chemical factories. They may have many uses.

The resin of very ancient trees, millions of years ago, has hardened into amber, a yellow jewel, found round the Baltic sea. Bronze Age men collected and traded it as far away as Crete, Egypt and Carthage, carrying it along the 'Amber routes' that have been tracked down by archaeologists.

Sometimes insects were trapped in the ancient resin, and formed beautiful fossils in the amber, with every little bristle showing clearly. Many of these 'amber fossils' were fakes, made from modern insects, and sold at a good profit.

Birds

This hole was made in partly rotten wood by woodpeckers, for nesting. The white streaky droppings patch marks the favourite perch of birds, just above. There is also a birds' nest, made of twigs, in the drawing below.

Witches' brooms

These big bunches of twigs are caused by an infection. Some of the stunted twigs can be used as cuttings and will grow into dwarf trees. There are tiny witches' brooms on some birch trees, caused by a mite.

Green flat patches

These green patches contain millions of microscopic plants. Under a microscope (high power) the bright green patch will show you cells dividing into twos and fours. The grey-green one is a mixture of green cells and fungus threads, living together. This mixture is called a lichen, and this one is a flat lichen. Why do these patches grow in some places only?

Lichens that are not flat

These can grow in different shapes and colours. They all contain green cells, as described above, but these are often hidden inside. They sometimes show up better when the lichen is wetted. Very few lichens are found where the air has acid smoke, unless they can grow on limestone rocks that neutralise the acid. The air is clean where you find several different kinds of lichen on trees and stones.

Mosses and ferns can also be found on some tree trunks or even small blackberry bushes, high up (*Autumn Trees,* page 25).

Streaks

The dark streak is caused by rain water running down always in the same place. Try wetting dry bark yourself. The pinkish streak has little soft blobs of pink yeast in it. The yeast is a microscopic plant that grows in sugar. The sugar is leaking out of the tree because it has been injured, perhaps by insects. Elms, vines and Sugar maple are well known for 'bleeding'. So we do not prune these trees in spring, when the sugary sap is rising up.

Bark beetles

These rows of small holes mark the exits of bark beetles. Their tunnels are between the inner surface of the bark and the outer surface of the wood, because they are eating the soft new tissues near the cambium. The female beetle ate out the wide burrow and laid eggs all along it. A small grub started out from each egg, making a widening burrow as it grew fatter and fatter. When the grub was full grown, it turned into a beetle and made its own exit hole. When the bark is peeled off, you can see that both bark and wood are grooved or *engraved. You* can engrave wax. If you use some smooth pale paper and rub dark wax crayon all over, you can make pale lines by engraving right through the wax with a metal or hard plastic point. If you use rough paper, or a broad scraper, you get different results.

Other insects

Many insects run up the tree bark, especially at night. Ants run to milk aphids. Spiders run to catch insects. Often you find cocoons, eggs, pupae, and cast skins in the bark. Birds often search the bark for insects.

Webs

Spider webs are very common on the furrowed bark of old trees. This one is the lace-web spider with its very sticky web. All over the trunk, if the light catches it, you can see the drag lines made by other spiders as they ran about. If the spider falls it is at the end of its own rope.

Toadstools, mosses

These are toadstools of the honey fungus, growing on a live tree. They will finally kill it. Birch trees often have a white shelf fungus which kills them in the end. But most of the toadstools that grow on dead wood or leaves have not killed the plant, and they are helping by decaying away the dead remains.

There are also many mosses and damp-loving animals on these rotting tree trunks. You will find them in the *Animals in the Soil* book.

Wood and annual layers

You can get a clean wooden stick, for cooking dough over a camp fire, by peeling a twig (page 32). The wood is usually more brittle than the bark you peel off, and it has a small pith in the middle. Elder often has a wide pith and you can peel it to make a doll that will not fall over, or poke it out to make a pipe. In the wood there are very narrow tubes, just big enough to see. These are used to carry water for the tree. In between the tubes are close-packed fibres, which are so fine that they look quite solid. But if you break the wood open, the fibres splinter apart. Planks are cut with the fibres going along them, and this is called along the grain. A plank cut across the grain would not be so strong.

Twigs that are more than a year old have *annual layers* of wood, and you can count them to find the age. Does this count agree with the girdle scars? If the layers are difficult to see, try cutting the twig on the slant. On a really big trunk you can find the year of your birth or famous dates in history, by counting in from the edge.

This is how the annual layers are made. The twig peeled easily, because of a slimy soft layer round the outside of the wood, called a cambium. The cambium makes the new wood layer for each year, adding it on outside the old. The new wood is always soft at first. In spring the cambium makes wood with many big tubes called early wood. During the summer, 'late wood' is made, and added in a layer outside the spring wood. Late wood usually has fewer and narrower tubes, with thick walls, and many fibres. It gives strength to the trunk.

In good sunny years a thick layer of wood will be made, but the layers can be thicker on one side of the trunk than the other.

When planks are cut down through the trunk, the annual layers makes lines going down, but they may be straight or wavy. Some of the tubes are cut lengthwise: use a lens. At the end of the plant, you will see the same annual layers, cut *across*. Try to work out where the plank was in the original log

Patterns of annual layers

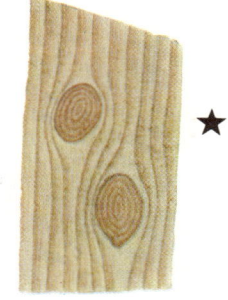

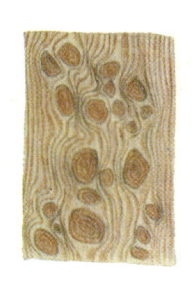

Saw a small log across in different ways: straight across, slanting, straight down, and you will see the different patterns made by the annual layers. Or cut a big beetroot. These make new layers every few weeks and may print on paper or clothes. Or make a model with layers of Plasticine, clay or coloured dough. Look at different pieces of solid wood to see how they were cut out of the trunk. Plywood or veneer is different (page 60). Wood makes good rubbings, if it is worn by the weather, or by scrubbing. The early wood wears down faster.

Where a tree forks you can find new patterns by cutting across with a saw. And if you have a piece of wood with horizontal layers, or a blancmange in layers, you can make patterns (contour lines) by digging pits and trenches in it. Try carving (page 57).

The centre of the trunk is the oldest wood, and it may decay, causing hollow trees. But some trunks have the central (heart) wood darkened with tannin, and this helps prevent decay. Oak has a very wide heart, and laburnum has a very dark one.

Where a side branch grows out, the annual layers still form round the trunk and round the branch. So the layers on the trunk gradually bury the lowest part of the branch. If you cut the wood through the buried part of a branch, you can see how the branch makes an oval *knot* in a plank, but the knot is tightly joined on. Knots with dark hearts can make a pretty pattern, and some kinds of wood, like burr-maple (page 45), have many small knots, side by side. If the trunk goes on forming layers round a *dead* side branch, the dead branch gets buried, bark and all, and never joins in firmly. This makes loose knots that often have resin (page 52) round them, and the timber is poor. Good furniture will not have knots. You need to look at builders' planks and packing-cases to make a collection. If side branches fall off while the tree trunk is still small, plenty of sound wood will form outside, with no knots. That is why the trunks of young forest trees are sometimes cleaned off, by pruning, nowadays.

55

Wood of different trees

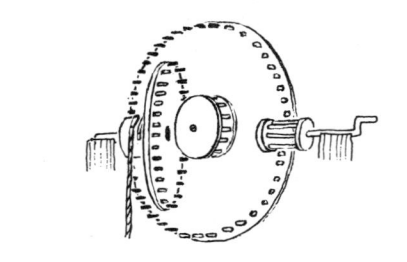

Early wood and late wood sometimes have different colours and show very clearly. The late wood of elm has a feathery look. But sometimes you can only see the annual rings by using a lens. Look for a band with many tubes, and a band with very few. Balsa is soft because it grows fast in the tropics, and it has poor annual layers because there are no seasons there.

The wood of conifers is often called *softwood*, but it is harder than the wood of some broad-leaved trees, such as balsa. Conifer wood does not have very clear layers and no really long tubes. Conifers grow fast and sell well, so a great many are planted nowadays. Slow-growing oak and hornbeam were more important in the time of Henry VIII, when oak was needed for ships, hornbeam for big gear wheels. We use metal now.

To get straight telegraph poles, the trees are grown close together. To get the curving beams for Tudor ships, people often bent the tree when it was young, or searched out curved pieces. Oak branches bend a lot.

Some trees have hard wood, some elastic, some very even, some resisting decay. Tree books and woodwork books will tell you about the different qualities and uses, and you can test some of them for yourself. If you look around you will see many uses of wood. Paper is made by separating wood or other plant fibres by machinery or chemicals. To get an idea of the method, mash up toilet paper in water, then dip a fine sieve into this mixture, drain off the water, and you have a layer of wet paper.

Rayon is made by dissolving the wood into a treacly solution, and forming fine filaments from this.

Testing wood

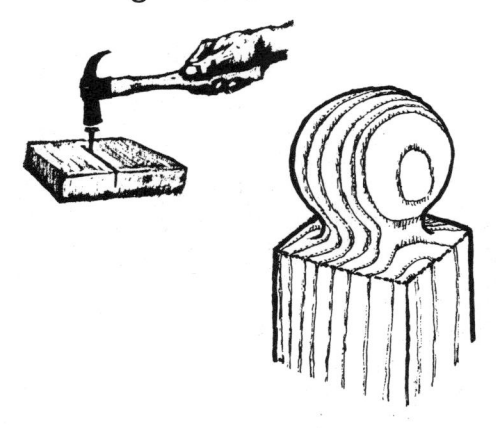

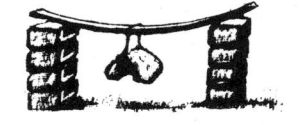

To test the strength of different woods you need pieces of the same size. But there are different kinds of strength. If you try with a nail and one hammer blow on each wood, you will test for one kind of *hardness* and also see if the wood tends to *split*.

Then try carving, and look for *even grain* — lime is good.

You can test resistance to *bending* or *breaking* by making your wood into a bridge, and hanging different weights from the middle. You need thin laths, about 2 cm by 2–3 mm, or narrow branches, but they must be at least 20 cm long. Even small laths need big weights to break them, which can be dangerous. Do the whole test very close to the ground, and use sandbags instead of metal weights, for sandbags do not bounce. Resistance to *compression* is harder to measure. Try laying the wood on the ground, placing a coin on it, and standing on this.

Brittleness can be tested by using equal thin planks or branches, and seeing how far you can bend them without breaking. You need a vice and must be very careful of the wood springing back. If the wood goes back to normal afterwards, and bounces back forcibly, it is *resilient*. Freshly cut twigs (green wood) are less brittle, more pliable than those that have dried out. What happens if you soak them again? Fresh branches are also resilient. Blacksmiths once had big hammers that pulled on a branch, as they came down, then were lifted up by the resilience ready for the next blow. Lathes were also made to work by the resilience of a green branch.

Planks are cut with the fibres (the grain) running along them. Is a small plank, cut across the grain, really weaker?

57

Warping and cracking

Freshly cut wood spits in a fire, because it contains a lot of water. Try weighing some fresh twigs, then dry them thoroughly and reweigh.

Wood shrinks as it dries, so the bark on a dry twig is too big for it, and goes into folds. Whole branches or tree trunks crack as they dry: look at old dry tree trunks or keep a piece of 10 cm branch in a warm room. The outside wood shrinks more than the rest and makes big cracks there. But if the trunk is cut in planks, they will usually bend or warp instead of cracking, and they can be levelled off with a plane.

If wood is made into furniture before it is properly dry, the furniture will go on shrinking and warp or crack. So the drying out, or *seasoning*, is important. The planks were once left for years with air circulating between them, but nowadays dry air is used, for speed, in a special building or kiln. Wood for dry climates or heated houses must be dried more. But if furniture is kept in a damp house, it may warp in the other direction. Wooden road blocks swell if they get flooded, and burst out of position. Measure a piece of seasoned wood, then soak it and watch for swelling and warping.

Polish

If you want to keep a piece of tree trunk with bark-beetle engravings, or a wood carving, it may crack as it dries. Oil or wax polish are used to slow the drying.

Wax polish also makes a smooth surface, and lets you see the grain clearly. Shellac and varnish do the same. Try these out.

What happens to the colour of wood that has been left a long time in the open? Does it make a difference if it was varnished or polished?

Rays

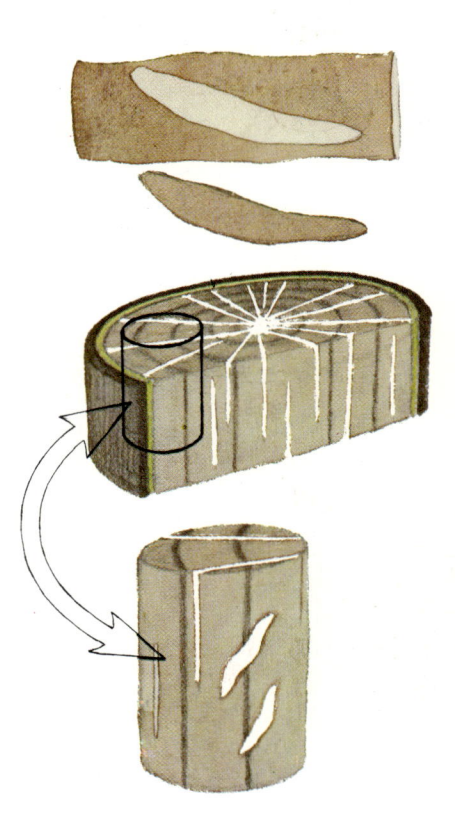

Polished oak furniture often has bright patches, called *rays*. And if you cut across any log or branch you will see narrow rays running into the centre: polishing helps.

Find the shaggy end of a block of balsa, where the fibres have been cut across roughly. Then smooth this with a sharp blade, to get a good cross section. The tubes (page 54) show up well, outlined in brown: use a lens. Down the sides you can see some of these tubes cut lengthwise. On the cross-section you can also see fine brown lines, nearly parallel. These are rays. Follow them to the edge of the wood and you will see them as short brown lines. The rays must be ribbon-shaped. The ribbon stands on its edge, and runs in from the edge of the trunk to the centre. Sometimes you can carve the wood to uncover a ray, but it is not easy. Oak wood is hard and the rays are wavy. Most trees have pale rays, not dark ones, but rather thick and strong, with no tubes in them. If oak is cut *along* the rays you often see bits of the wide ribbon. If it is cut across the rays, they look like short lines.

A cylindrical chair-leg can have flashes of medullary ray on two sides but only narrow lines on the space between. The annual layers go on as usual between the rays, so the rays run right across the spring and summer wood.

A little brown line might be a ray cut across or it might be a tube, cut along its length for a short way. But a tube should have a hollow, if you look with a lens. Also, tubes are more common in spring wood, so tubes are always arranged in layers of some kind. But rays run right across spring wood and late wood, and are found crossing both, equally. Some foreign woods have quite big rays even when they are cut *across*. The rays of cedarwood and plane (lace-wood) are more obvious than the annual layers. Lace-wood was used for some London underground trains.

Plywood and veneers

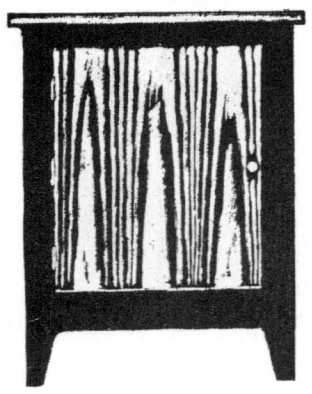

Three-ply wood is made of three layers with the grain going different ways. See if you can make this out with a broken piece. If the layers are stuck with modern resin glues the result is stronger than solid wood, and less likely to warp. Try making some plywood with the thin wood from a matchbox.

Plywood can be seven-ply, or three-ply. There are many kinds. Look at hardboard, chipboard, blockboard and softboard. How do you think they are made?

To get big sheets of plywood, a whole log is twisted round and round against a cutting blade, so that the wood comes off like unrolling a Swiss roll, but much thinner. Try this with a short piece of carrot that has dried out a little. All this wood is cut across the medullary rays, so over the whole area you can see tiny medullary rays but not big flashes. The top layer is often made of a more expensive wood, and this is called a veneer.

The annual layers make patterns along a big sheet of veneer. If there was a bulge in the layers in one place, each time the knife came round it would cut through it, so there is a repeating pattern, just a bit different, each time. You can sometimes see this repeating pattern in modern halls lined with wood.

Veneers can be puzzling. This small table top has tiny cross sections of medullary rays all over it, even going round the curve at the edge. It is veneered plywood with veneer stuck all round the edges as well as on top. Veneers that show flashes of medullary rays are not cut Swiss-roll-fashion but must be cut along a radius. Therefore they are much more expensive. Symmetrical veneers are also used on some furniture. Can you work out how they could be cut?

Different places, different plants

Try to find three beech trees, and see what is growing under them. Then compare with three oak trees or other trees. In a park, these things are planted but in the country you will find there are natural differences. Very few plants will grow under beech or pine trees.

Try to see what causes the difference. Is the soil the same, with the same layers? Is the soil the same all the year? Some plants cannot stand being waterlogged in winter. Can you find out how much water there is in a kilo of each soil?

Some people think that beech leaves and pine needles have chemicals that poison other plants. Could you test this out?

Is there the same amount of light under the different trees? To measure light, lay a bed sheet on the ground and hold a photographic light meter about 1 metre above it, facing down. Then try with the other tree. If you try in different places under the tree you can map where the light is weakest. Is it the same all day, and all year? Suppose you measured one tree when the sun was out, and the other when it was behind a cloud?

If you can find a hedge that runs east and west, look at the sunny side and the shady side, and see if the plants and animals are different.

When you go to different places, such as chalky places, windy places, hills and valleys, look at the trees and see if there are different kinds, or if they grow differently. Sometimes there are specially young or old trees that are worth looking at.

Dandelions and other plants can grow in light or shade, but look different. Light affects many plants.

Some plants have weak stems but climb up others. How do they do this?

Roots

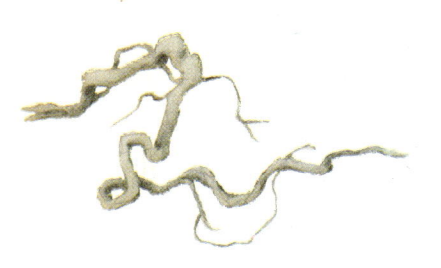

When soil is dug away, try to see the roots of trees, how far they go, and how they twist round stones in hard ground. If a tree has blown over, try to see why. Look for the smallest branches of the roots. How far do you think they would stretch if they were all laid end to end? Roots help to hold the tree up but they also take in water and mineral fertiliser for the leaves and branches.

Dead leaves and tree trunks are eaten by woodlice and insects. You will often find these under the bark or in leaf litter. Bacteria and fungi also decay tree trunks and leaves. Try to find the stages. Toadstools grow on the trunks they are decaying. Mosses and lichens also like decaying tree trunks and leaves because they are moist and they give out mineral fertiliser.

Some leaves decay slowly. The dead leaves under a beech tree may last several years, and all this time the tree cannot get at the minerals. But if you search in the upper layers of leaf litter, where it is damp, you may find special fat roots that have grown *upwards* into it. These roots have a layer of fungus round them. You can see strands if you put them in water and use a powerful lens. The fungus runs across from the tree roots into the dead leaves, and it seems to get the minerals out of the dead leaves and pass them on to the tree roots. The fungus probably gets some sugar from the tree roots. In autumn, the fungus sends up its own toadstools with dusty spores, and the tree forms its own seeds. Some trees have special toadstools. The birch has the red one with white spots, called Fly agaric.

Of course there are many other fungi that do not coat tree roots. Some fungi make leaves and trunks decay. Some attack and kill live trees.

Nature trails

When you have found out some interesting things about the trees in your park, you can show them to other people by setting up a nature trail. Choose the first interesting place, and knock a stake into the ground, numbered '1'. Then make an arrow pointing to the next interesting place, labelled '2'. At each place you can put up a notice or lay out a picture or specimens to show the interesting things. If you do this in a park, you will need to ask permission and you cannot usually knock in stakes or set up notices. Instead, you can make out a map and guide to give to visitors.

Moving pictures

If you can borrow a film camera, and clamp it up, about 120 cm above the ground, and facing downwards, you can make a film. You will need help from an adult, and perhaps some extra lights. Suppose you have watched a sycamore fruit growing and know the stages. Lay on the ground, in view of the camera, a piece of Plasticine to represent the stalk. Keep this fixed. Then make a model of the very young sycamore in Plasticine or cut-out paper, and put it on the stalk. Run the camera for a second or less. Then leave the stalk in position, but put on the next size of sycamore. Run the camera again, and so on. If you want titles you will have to write them out first, to the right size. You can try all sorts of things, like buds opening and tree trunks growing. If you have no camera you can make a drawing seem to move by using a thick notebook. Draw the first stage on the corner of the back page, the next on the same corner of the next page, and so on. Then run them through your fingers.

Index

a
acorns   29
anglemarks   42
annual layers   54 55
aucuba flowers   27 28

b
bark   32 on
bark beetles   53
bast   32
bees   27
buds   4 5 6 8
bud scales   4
bumps   46
burrs   46

c
cambium   3 54
catkins   24 25
cork, cork oak   33

d
decay   62
double flowers   28

e
engraving   53
evergreens   11

f
falling from trees   13
films   63
flowers   24
food   10
forcing   4
fruits   26
furrowed bark   47 50

g
germination   29
girdle scars   5 37
grain   54
growth curve   14
growth of stems   14
growth series   26

h
height of tree   18
holly flowers   28
honey guides   24

i
insects   24

k
knots   55

l
lammas shoots   16
leaf, abnormal   12
       colours   9
       evergreen   11
       kinds   12
       mosaics   10
       old   9
       series   6
       symmetrical   12
       use   10
       wings   7
       young   8 9
lenticels   33 38 39
lichen   52
light   61

m
moving pictures   63

n
nature trails   63
nectar   24

o
oak   21

p
paper   56
petals   24
pistil   24
pith   54
plaster casts   12
pollen   24 27 28

pollen trees   28
pollination   27
polish   58
pruning   22 23

r
rays   59
resin   51 52
roots   10 62

s
scale insect   37
scent   24
seals   43
seeds, seedlings, seed
    leaves   29 30–31
seed trees   29
self pollination   28
size of trees   18
soil   61
spread of tree   18
spring   4
stamens   24
stems   14
stopping growth   15
suckers   46

t
tannin   33
toadstools   53 62
twigs   17 38
twig scar   13 37

v
veneers   60

w
wood   54 on
wrinkles   42